DIGRAPHS: THEORY AND TECHNIQUES

DIGRAPHS: THEORY AND TECHNIQUES

D. F. Robinson
*University of Canterbury, Christchurch,
New Zealand*

L. R. Foulds
*Massey University, Palmerston North,
New Zealand*

GORDON AND BREACH SCIENCE PUBLISHERS
New York   London   Paris

Copyright © 1980 by Gordon and Breach, Science Publishers, Inc.

Gordon and Breach, Science Publishers, Inc.
One Park Avenue
New York, NY10016

Gordon and Breach Science Publishers, Ltd
42 William IV Street
London WC2N 4DE

Gordon & Breach
7–9 rue Emile Dubois
Paris 75014

Library of Congress Cataloging in Publication Data
Robinson, David Foster, 1936—
   Digraphs: theory and techniques.

   Bibliography: p.
   Includes index.
   1. Directed graphs. I. Foulds, L.R., 1948—joint author. II. Title.
QA166.15.R62   511'.5   79-7466

This book is gratefully dedicated to our wives Juliet and Maureen

PREFACE

Since graph theory has become a systematic tool in
recent years, there has been a rapid growth in the study
of its theory and applications. A great many of the real-
life applications of graph theory are really concerned
with directed graphs, or as we prefer to call them,
digraphs. There are good reasons for this. Many real-
world systems can be characterized by a binary relation
between elements. Since this is what a digraph is, the
theory can often be applied to analyse problems in such
fields as traffic and electrical engineering,
communications, operations research, logic, sociology,
and linguistics.

There does not appear to exist in print a text book
designed to present the theory and applications of
digraphs at an elementary level. Thus this book has been
written for first or second year university students (or
senior high school students) studying mathematics,
engineering, business or the social or biological sciences
who wish to gain an introductory knowledge of digraphs.
The book assumes very little mathematical knowledge.
Most of the mathematical necessities are included in the
preliminary chapter, making the book virtually self-
contained. A knowledge of the elementary properties of
matrices is required for chapter 8. If the teacher finds
this knowledge is lacking in his students and he is
unwilling to devote the short time necessary for its
presentation, the chapter could be conveniently omitted
without loss of continuity.

The book has arisen out of the experience of
teaching the material to first year mathematics students
at the University of Canterbury and Massey University.
A feature of the book is the inclusion of applications in
each chapter as the theory is developed. We have sought
these applications from diverse areas. We believe it is

important to let the student know very early that the theory of digraphs can be applied to many branches of knowledge. To miss the applications of digraphs is to miss a major part of their importance. We hope that students will be stimulated to develop new applications in their own fields of interest.

Chapter 1 is basically an introduction to digraphs and their elementary properties. We have included an application to logic. Chapter 2 is concerned with the structure of digraphs, particularly the component structure. Many of the exercises indicate applications. Chapter 3 introduces acyclic digraphs and applications to logical numbering and problems of precedence. Chapter 4 introduces tournaments. Although this is an everyday topic this chapter is reasonably theoretical. We have introduced different types of tournaments and their structure from a mathematical point of view. Chapter 5, on rooted trees, contains a lengthy section on the integer programming technique of branch and bound enumeration. In operations research many of the applications of digraphs are concerned with discrete optimization. Hence this chapter, and the next two on networks, are designed to give the reader a brief introduction to the relationship between digraphs and one of their major application areas - Operations Research. Chapter 8 is a more theoretical development on some of the matrices associated with digraphs. No book would be complete without a major reference to the large class of digraphs which are symmetric, and are usually called graphs. In Chapter 9 some of the basic concepts of graphs are introduced including an application to finding minimal spanning trees. This chapter has been placed last as the purpose of the book is to present the theory and application of digraphs not graphs, which have been treated fully elsewhere.

The chapters: 0, 1, 2, 3, and 8, with a limited introduction to matrices in chapter 0 would provide a one-semester course on the theoretical aspects of digraphs. The other chapters are more applied and are self-contained. They could be added or substituted to provide a course with a more applied emphasis. The following plan outlines these ideas.

The book contains a large number of exercises. These range in difficulty from the drill type to unsolved problems from the research literature. The reader is strongly encouraged to attempt as many of the exercises as possible. In order to appreciate the thrills of mathematics, as with music, the student must be prepared

Preface

to work hard in trying the ideas out for himself.
Mathematics is not a spectator sport! Solutions to
selected exercises are provided at the back of the book.
The more difficult problems are starred.

As this book is designed to merely introduce the
theory of digraphs we have provided a guide to the
literature for future study on the subject as an appendix.

The authors are grateful for this opportunity to
express their thanks for the support of the Mathematics
departments of Canterbury and Massey Universities which
they enjoyed while writing this book.

D.F.R.

L.R.F.

Palmerston North

CONTENTS

Plan of the Book   xv

0. Preliminaries

   0.1   Set Theory   1
   0.2   Functions   3
   0.3   Summation   4
   0.4   Linear Algebra   5
   0.5   Maximum and Minimum   10
   0.6   Mathematical Induction   11

1. Basic Concepts of Digraphs

   1.1   Relations   13
   1.2   Equivalence Relations   16
   1.3   Exercises   17
   1.4   Digraphs   18
   1.5   Indegree and Outdegree   19
   1.6   Exercises   22
   1.7   Walks   24
   1.8   Joining and splitting walks   26
   1.9   An application to logic   29
   1.10 Exercises   32
   1.11 Subdigraphs and partial digraphs   34
   1.12 Converse   36
   1.13 Isomorphism   38
   1.14 Exercises   41

2. Digraph Structure

2.1	Structure	43
2.2	Reachability and Equivalence	44
2.3	Component analysis	48
2.4	Exercises	51
2.5	Condensation	53
2.6	Components and Closed Walks	56
2.7	Weak components	59
2.8	Exercises	60
2.9	Strong Digraphs	63
2.10	Distance Calculations	64
2.11	Exercises	67

3. Acyclic Digraphs

3.1	Definitions and Tests	70
3.2	Applications of logical numbering	75
3.3	Levels	79
3.4	Exercises	80
3.5	Essential and Inessential Arcs	81
3.6	Partial Orderings (non-linear rankings)	86
3.7	Linearisation of Orderings	91
3.8	Exercises	92

4. Tournaments

4.1	Introduction	95
4.2	Unilateral Digraphs	98
4.3	Tournament Structure	101
4.4	Strong Tournaments	102
4.5	Acyclic Tournaments	105
4.6	Exercises	106
4.7	Outdegree Analysis	108
4.8	Tournaments with ties	112
4.9	Exercises	113

5. Rooted Trees

5.1	Introduction	115
5.2	Applications of Rooted Trees	116
5.3	Exercises	121
5.4	Branch and Bound Enumeration	124
5.5	Rooted Trees and Games	132
5.6	Exercises	134

Contents

6. Networks

6.1	Introduction	138
6.2	Shortest Paths	140
6.3	Flowcharts	144
6.4	Exercises	148

7. Activity Networks

7.1	Introduction	151
7.2	Constructing the Network	153
7.3	The Critical Path Method	156
7.4	The Activity-arc Approach	165
7.5	Exercises	167

8. Matrices of Digraphs

8.1	Introduction	169
8.2	The Adjacency Matrix	169
8.3	Digraph Properties and Matrix Properties	173
8.4	The Incidence Matrix	176
8.5	Exercises	178
8.6	The Cycle Matrix	182
8.7	The Fundamental Cycle Matrix	187
8.8	The Fundamental Cut-Set Matrix	192

9. Graphs

9.1	The Relationship between Graphs and Digraphs	201
9.2	Subgraphs and Isomorphism	203
9.3	Complete, Complementary and Self-complementary graphs	205
9.4	Exercises	207
9.5	Planar and Bipartite Graphs	212
9.6	Trees	213
9.7	Exercises	222

Solutions to Selected Problems	226

Solutions to selected problems:

Chapter 1   226
2   228
3   231
4   233
5   234
6   238
7   239
8   241
9   245

Further Reading   250

Index   253

PLAN OF THE BOOK

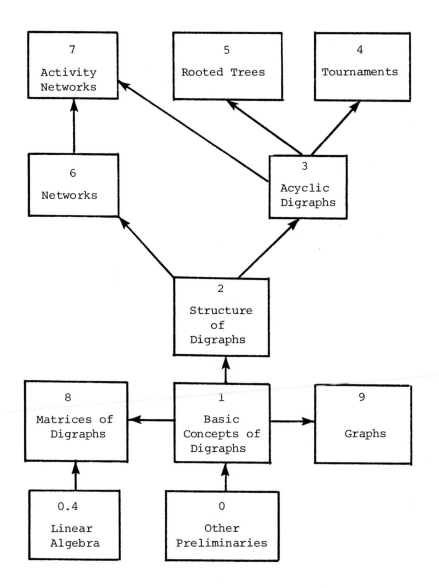

CHAPTER ZERO

PRELIMINARIES

0.1 Set Theory

One of the basic ideas of all mathematics is that of
a set, and this book uses the notation of sets to describe
many of the concepts we will meet. A set is a collection,
pile or heap of objects, mathematical structures, numbers,
people, words or anything else. All that matters is that
for any set any object either is or is not an element of
the set.

For a small set we may list the elements: the set
$\{1,2,3\}$ consists of the numbers $1,2,3$: each is an element
of the set. We write

$$2 \in \{1,2,3\},$$

to mean '2 is an element of the set whose elements are
1, 2 and 3. On the other hand 4 is not, so we write
$4 \notin \{1,2,3\}$.

The order of writing the elements is not important,
$\{3,1,2\}$ is the same set. On the other hand the style of
brackets, '$\{,\}$' is important.

A set can also be described by the property
belonging precisely to the elements of the set:

$$\{3,4,5\} = \{x: (x - 2) \in \{1,2,3\}\}.$$

We shall need to combine sets in four ways to define
new sets. Most importantly we need a set which consists
of those elements which are elements of both of a pair of
sets. Thus 3 and 4 are the only elements common to
$\{1,2,3,4\}$ and $\{3,4,5\}$.

We write $\{1,2,3,4\} \cap \{3,4,5\} = \{3,4\}$. This is the
intersection operation. Formally:

$$A \cap B = \{x: (x \in A) \text{ and } (x \in B)\}.$$

Sometimes two sets have no elements in common. Thus we need a set, the empty or null set, which has no members. We write $\emptyset$ for this.

$$\{1,2\} \cap \{4,6\} = \emptyset$$

We also need to combine sets to give those elements which belong to one set or the other. This gives the union:

$$\{1,2,3,4\} \cup \{3,4,5\} = \{1,2,3,4,5\}.$$

$$A \cup B = \{x: (x \in A) \text{ or } (x \in B)\}.$$

Notice that we include those elements which belong to both sets.

We next need to compare two sets. Two sets are equal if they have exactly the same elements. A set A is contained in set B if every element of A is also an element of B. Thus $\{1,3\}$ is contained in $\{1,3,5\}$. We write

$$\{1,3\} \subseteq \{1,3,5\}$$

We have other forms of words that convey the same information:

$$\{1,3,5\} \text{ contains } \{1,3\},$$

$$\{1,3\} \text{ is a subset of } \{1,3,5\},$$

$$\{1,3,5\} \text{ is a superset of } \{1,3\}.$$

We consider the empty set to be contained in every set. Every set also contains itself.

When every element of set A is also an element of set B we are often interested in those elements of B which are not elements of A. We write

$$B \setminus A = \{x: (x \in B) \text{ and } (x \notin A)\},$$

and call $B \setminus A$ the complement of A relative to B. For example

$$\{1,2,3,4,5\} \setminus \{2,4\} = \{1,3,5\}.$$

The fourth method of combining sets forms ordered pairs. The reader may be familiar with the use of ordered pairs (x,y), where x and y are real numbers, to represent a point in the plane relative to given axes. A comparable use of ordered pairs is to assign to each person his height in centimetres and weight in kilograms. A person 180 cms. tall and weighing 72 kg. would then be

recorded as (180,72). A person whose ordered pair was
(72,180) would be quite a different shape!
    In general if A and B are any two sets, A × B, called
the cartesian product (after René Descartes, who founded
co-ordinate geometry) is the set of all ordered pairs
(a,b) with a ∈ A and b ∈ B.

$$A \times B = \{(a,b) : (a \in A) \text{ and } (b \in B)\}$$

Thus $\{1,2\} \times \{2,3,4\} = \{(1,2),(1,3),(1,4),(2,2),(2,3),(2,4)\}$
Two ordered pairs (a,b) and (c,d) are equal if and only if
a = c and b = d.
    We shall sometimes be concerned with the number of
elements in a set. This is called its cardinality, and we
write |A| for the cardinality of A. Then |{0,1,2}| = 3
and |∅| = 0.
    A set like the set of all books in the world at a
particular instant has a large number of elements and it
would be impossible to determine its exact cardinality.
But there are certainly not more than 10^{17} books -
approximately the number of standard-sized paper-backs to
cover the whole surface of the earth, standing upright.
The set of books is thus finite . On the other hand the
set of all natural numbers {1,2,3,4, ...} has more elements
than any number we can choose. The set of natural numbers
is then called infinite.

0.2 Functions

    It is very common in mathematics to wish to
associate elements of one set with elements of another.
One association between people and numbers is to assign to
each person his year of birth. Each person has a single
number; many people are associated with the same number,
being born in the same year. At the time of writing
no-one has the number of 2000. We can also set up an
association between books and numbers: each book has a
certain number of characters printed in it. A
correspondence between English words and the alphabet comes
from assigning to each word its first letter.
    All of these are examples of functions. A function is
a rule by which to each element of a set A we assign
exactly one element of the set B. If the function is
denoted by f, and x ∈ A, we write f(x) for the element of
B associated with x. Each element of B can be associated
with many elements of A, or just one, or no element
of A.
    If A is a finite set we can define a function not by a

formula in words or symbols, but by stating explicitly the
value of f(x) for each x in A.
     Suppose A = {0,1,2,3} and B = {1,3,5,7}.
Then we can define f by, for example,

$$f(0) = 3, \quad f(1) = 7, \quad f(2) = 3, \quad f(3) = 7$$

Another function from A to B is

$$g(0) = 7, \quad g(1) = 3, \quad g(2) = 1, \quad g(3) = 5$$

This has the property, which f does not, that each element
of B occurs as the element associated with exactly one
element of A.
     We describe this by saying that g is an invertible
function (or one-one correspondence) from A to B.   In each
such case there is another function h also invertible,
from B to A with the property that

$$h(y) = x \text{ if and only if } g(x) = y.$$

In this case

$$h(1) = 2, \quad h(3) = 1. \quad h(5) = 3, \quad h(7) = 0.$$

The function h is called the inverse of G.   An attempt to
construct an inverse for f should show the reader why it is
impossible.

0.3 Summation

     It is necessary in many places in this book to add a
series of numbers.   We shall for convenience use the
summation shorthand notation.   We use the Greek letter
Σ (sigma) to indicate summation of all the terms of
the type following the sigma, whose suffix fulfils
conditions written below, and in some cases above, the
sigma.
     Thus if S = {1,3,5,7},

$$\sum_{i \in S} i^2$$

means the sum of the squares of all elements in the set S.
Hence

$$\sum_{i \in S} i^2 = 1^2 + 3^2 + 5^2 + 7^2 = 84$$

     Very often the set is the set of all integers
between two integers a and b, (including both a and b).

A special notation

$$\sum_{i=a}^{b}$$

is used for this. For example,

$$\sum_{i=1}^{5} (2i - 1) = 1 + 3 + 5 + 7 + 9 = 25$$

0.4 Linear Algebra

(This section is required only for Chapter 8, and we introduce here only those concepts of linear algebra actually used in that chapter).

A matrix is a rectangular array of real numbers, such as

$$\begin{bmatrix} -2 & 3.5 & 4 & 9 \\ -0.5 & 2 & 3 & 8 \\ 5 & 6 & 7 & 0 \end{bmatrix}$$

The numbers are set out in horizontal rows such as

$$-2 \quad 3.5 \quad 4 \quad 9,$$

and vertical columns such as

$$-2$$
$$-0.5$$
$$5$$

A matrix may have any positive integral number of rows, and any positive integral number of columns. If there are m rows and n columns the matrix is described as an m × n matrix. The example is a 3 × 4 matrix. A matrix with equal numbers of rows and columns is called square.

The numbers in the matrix A are called the entries of A. For reference, each entry is identified by the row and column in which it appears: we give the symbol a_{ij} to the entry in the ith row and jth column of A. The row and column suffices are thus an ordered pair. Rows are numbered from the top and columns from the left. Thus if the example matrix is A,

$$a_{11} = -2, \ a_{12} = 3.5, \ a_{21} = -0.5, \ a_{34} = 0.$$

Where necessary we will write $A_{m \times n}$ to indicate that
A is an m × n matrix: in definitions reference to the
number of rows and columns is necessary, but in
applications it is determined by the context. We write
$$A_{m \times n} = B_{p \times q},$$ if and only if

(i) m = p

(ii) n = q,
so that they have the same numbers of rows and
columns, and

(iii) $a_{ij} = b_{ij}$; $1 \leqslant i \leqslant m$; $1 \leqslant j \leqslant n$, so that
corresponding entries are equal also.

Two matrices may be added only if they have the same
numbers of rows and columns, and then
$$A_{m \times n} + B_{m \times n} = C_{m \times n},$$
where $c_{ij} = a_{ij} + b_{ij}$, $1 \leqslant i \leqslant m$, $1 \leqslant j \leqslant n$.
For example

$$\begin{bmatrix} 1 & 0 & 2 \\ 0 & 1 & -2 \end{bmatrix} + \begin{bmatrix} 0 & 1 & 3 \\ -1 & -1 & 4 \end{bmatrix} = \begin{bmatrix} 1 & 1 & 5 \\ -1 & 0 & 2 \end{bmatrix}$$

The m × n matrix all of whose entries are zero is
called the zero m × n matrix and written $0_{m \times n}$, or simply
0, if the context is not ambiguous.

Some matrices have only a single row or column: a
1 × n matrix is called a row vector; an m × 1 matrix is
called a column vector, and n,m respectively will be
called the dimension of the vector.

Any matrix can be multiplied by a real number. If
k is any real number and $A_{m \times n}$ a matrix, then the real
multiple $kA_{m \times n}$ is
$$kA_{m \times n} = B_{m \times n},$$
$$ka_{ij} = b_{ij}, \quad 1 \leqslant i \leqslant m, \ 1 \leqslant j \leqslant n.$$
Thus

$$3 \begin{bmatrix} 1 & 0 & 1 \\ 0 & 1 & -2 \end{bmatrix} = \begin{bmatrix} 3 & 0 & 3 \\ 0 & 3 & -6 \end{bmatrix}$$

Real numbers are often called scalars in this context,
and the real multiple is better known as the scalar
multiple.

If $A_{1 \times n}$ is a row vector and $B_{n \times 1}$ is a column vector

of the same dimension we can define a <u>scalar product</u> or
<u>inner product</u>

$$A_{1 \times n} \cdot B_{n \times 1}$$

by multiplying together the corresponding elements and
adding.

Thus if $A = \begin{bmatrix} 0 & 1 & 2 & 3 \end{bmatrix}$

and $B = \begin{bmatrix} 2 \\ 0 \\ 4 \\ 3 \end{bmatrix}$, we calculate

$$A.B = \begin{bmatrix} 0 & 1 & 2 & 3 \end{bmatrix} \begin{bmatrix} 2 \\ 0 \\ 4 \\ 3 \end{bmatrix} = 0 \times 2 + 1 \times 0 + 2 \times 4 + 3 \times 3 = 17.$$

In general

$$A_{1 \times n} \cdot B_{n \times 1} = \begin{bmatrix} a_{11} & a_{12} \cdots a_{1n} \end{bmatrix} \cdot \begin{bmatrix} b_{11} \\ b_{21} \\ \cdot \\ \cdot \\ b_{n1} \end{bmatrix} = a_{11}b_{11} + a_{12}b_{21} + \cdots a_{1n}b_{n1}$$

using the summation notation,

$$A_{1 \times n} \cdot B_{n \times 1} = \sum_{j=1}^{n} a_{j}b_{j}$$

Using this inner product we define the <u>product</u> of
two matrices. As with addition, we cannot always multiply
two matrices; the product $A_{m \times n} B_{p \times q}$ is defined only if
$n = p$, and then

$$A_{m \times n} \ B_{n \times q} = C_{m \times q}$$

The entries of C are defined in terms of the rows of
A and the columns of B, each of which is a vector of
dimension n. Then if A_i now stands for the ith row of
A, and B_j for the jth column of B, the entry c_{ij} is the
inner product of A_i with B_j.

Thus if $A_{2\times3}$ = $\begin{bmatrix} 2 & 0 & 10 \\ 6 & 1 & 2 \end{bmatrix}$

and $B_{3\times4}$ = $\begin{bmatrix} 1 & 1 & 2 & 3 \\ 4 & 4 & 0 & 1 \\ 8 & 6 & 2 & 1 \end{bmatrix}$,

the product AB is defined and will be a 2 × 4 matrix.

A_1 = [2 0 10]   and B_1 = $\begin{bmatrix} 1 \\ 4 \\ 8 \end{bmatrix}$

so c_{11} = $A_1 \cdot B_1$ = [2 0 10] . $\begin{bmatrix} 1 \\ 4 \\ 8 \end{bmatrix}$ = 82;

c_{12} = $A_1 \cdot B_2$ = [2 0 10] . $\begin{bmatrix} 1 \\ 4 \\ 6 \end{bmatrix}$ = 62.

Calculating the other entries similarly,

AB = C = $\begin{bmatrix} 82 & 62 & 24 & 16 \\ 26 & 22 & 16 & 21 \end{bmatrix}$

In general, if $C_{m\times q}$ = $A_{m\times n}$ $B_{n\times q}$

$$C = \begin{bmatrix} A_1 \cdot B_1 & A_1 \cdot B_2 & \cdots & A_1 \cdot B_q \\ A_2 \cdot B_1 & A_2 \cdot B_2 & \cdots & A_2 \cdot B_q \\ \cdot & & & \cdot \\ \cdot & & & \cdot \\ \cdot & & & \cdot \\ A_m \cdot B_1 & A_m \cdot B_2 & \cdots & A_m \cdot B_q \end{bmatrix}$$

In terms of the entries,

$$c_{ij} = a_{i1} b_{1j} + a_{i2} b_{2j} + \cdots + a_{in} b_{nj}$$

$$= \sum_{k=1}^{n} a_{ik} b_{kj}$$

It may be noticed that if m = 1 and q = 1, then

$A_{1 \times n} . B_{n \times 1}$ is a 1 × 1 matrix whose entry is $A_{1 \times n} . B_{n \times 1}$.

If A and B are two matrices such that AB is defined, BA may still not be defined; both are defined only if A is m × n and B is n × m; then AB is m × m and BA is n × n, so that AB = BA is possible only if A and B are both square. Even then, generally, AB ≠ BA. For example

$$\begin{bmatrix} 1 & 2 \\ 1 & 0 \end{bmatrix} \begin{bmatrix} 1 & -1 \\ 1 & -1 \end{bmatrix} = \begin{bmatrix} 3 & -3 \\ 1 & -1 \end{bmatrix}$$

but $\begin{bmatrix} 1 & -1 \\ 1 & -1 \end{bmatrix} \begin{bmatrix} 1 & 2 \\ 1 & 0 \end{bmatrix} = \begin{bmatrix} 0 & 2 \\ 0 & 2 \end{bmatrix}$

Just as 1 is the special number with the property 1.x = x for every real number x, so for any positive integer n there is a square matrix $I_{n \times n}$ (or simply I) such that for any $A_{m \times n}$:

$$I_{m \times m} A_{m \times n} = A_{m \times n}; \quad A_{m \times n} I_{n \times n} = A_{m \times n}$$

The form of this matrix is

$$I_{m \times n} = \begin{bmatrix} 1 & 0 & 0 & 0 & \dots & 0 \\ 0 & 1 & 0 & 0 & \dots & 0 \\ 0 & 0 & 1 & & \dots & 0 \\ . & . & . & & & \\ . & . & . & & & \\ . & . & . & & & \\ 0 & 0 & 0 & & \dots & 1 \end{bmatrix}$$

The entries are zero except on the <u>leading diagonal</u>, where they are unity. $I_{n \times n}$ is called the (n by n) <u>identity</u> or <u>unit matrix</u>.

If A is a square matrix we can find AA, or A^2. Then we can define $A^3 = AA^2$, ... $A^m = AA^{m-1}$, for any positive integer m. Since matrix multiplication is associative, when it is defined, we also have $A^3 = A^2A$, $A^4 = (A^2)^2$ and so on.

When we defined the inner product of two vectors earlier, the left vector was a row vector, and the right a column vector. The operation as we defined it cannot be used to multiply together two row vectors or two column vectors.

If $A = [a_{11}, a_{12}, \ldots, a_{1n}]$ and
$V = [b_{11}, b_{12}, \ldots, b_{1n}]$

are two row vectors and we wish to use the expression
$a_{11} b_{11} + a_{12} b_{12} + \ldots + a_{1n} b_{1n}$, it is necessary to turn
B into a column vector

$$B^T = \begin{bmatrix} b'_{11} \\ b'_{21} \\ \cdot \\ \cdot \\ b'_{n1} \end{bmatrix}$$

in which $b'_{i1} = b_{1i}$. Then B^T is called the <u>transpose</u> of
the vector B. The transpose operation can be applied
to any matrix, interchanging the rows and columns, so that
if A is m × n then A^T is n × m and if it has general
entry a'_{ij},

$$a'_{ij} = a_{ji} \qquad \begin{array}{l} i = 1, \ldots, n \\ j = 1, \ldots, m \end{array}$$

For example

$$\begin{bmatrix} 1 & 4 & 9 & 4 \\ 2 & 1 & 3 & 0 \\ 3 & 6 & 8 & 1 \end{bmatrix}^T = \begin{bmatrix} 1 & 2 & 3 \\ 4 & 1 & 6 \\ 9 & 3 & 8 \\ 4 & 0 & 1 \end{bmatrix}$$

If A is square, then A and A^T will have the same numbers
of rows and columns and identical diagonal entries. It
is possible that $A = A^T$ (as, in particular, for $I_{n \times n}$,
$0_{n \times n}$). Such a matrix is called <u>symmetric.</u> A matrix is
symmetric if and only if it is square and

$$a_{ij} = a_{ji} \qquad \begin{array}{l} i = 1, \ldots, m \\ h = 1, \ldots, m. \end{array}$$

0.5 Maximum and Minimum

It is necessary in many places to identify the
element of a set of real numbers which is the smallest.
This element is called the <u>minimum</u>. We write

min (S)

in such cases. Sometimes the numbers under consideration
are the values taken by the elements of a set S under some

real-valued function. Then we may write either

$$\min \{f(x): x \in S\} \text{ or } \min_{x \in S} \{f(x)\}.$$

For example, if $S = \{2,4,6,8\}$

$$\min (S) = \min \{2,4,6,8\} = 2$$

$$\min_{x \in S} \{(x-3)^2\} = \min \{(x-3)^2 : x \in S\} = 1.$$

Note that in the latter example the minimum is attained
at two elements.

In the same way the largest element of a set of real
numbers is called its __maximum__ and we write

$$\max \{2,4,6,8\} = 8$$

$$\max_{x \in S} \{(x-3)^2\} = \max \{(x-3)^2 : x \in S\} = 25.$$

0.6 Mathematical Induction

Several of the theorems in this book are the
assertions that some statement is true about all natural
numbers. Such theorems are formally proved by
__mathematical induction__. We begin the explanation of this
technique by an example. Suppose we find the sum of the
first n natural numbers:

$$n = 1: \quad 1 = 1 \qquad\qquad = 1.(1 + 1)/2$$

$$n = 2: \quad 1 + 2 = 3 \qquad\qquad = 2.(2 + 1)/2$$

$$n = 3: \quad 1 + 2 + 3 = 6 \qquad\qquad = 3.(3 + 1)/2$$

$$n = 4: \quad 1 + 2 + 3 + 4 = 10 = 4.(4 + 1)/2$$

Looking at the numbers $1,3,6,10$, and the continuation
$15,21,28,36, \ldots$ we may observe that they appear to obey
the pattern:

$$S_n = 1 + 2 + \ldots + n = \sum_{i=1}^{n} i = n(n + 1)/2,$$

as shown by the expressions on the far right above. The
statement is certainly true as far as we have gone, but is
it true for $n = 1000$ or $1,000,000$? The labour involved
in checking directly would be considerable and still leave
the problem undecided even for $1,000,001$.

Instead we prove that S_n does equal $n(n + 1)/2$ for all
natural numbers n by using the principle of mathematical
induction.

Principle of Mathematical Induction

Let $P(n)$ be a statement about the natural number n.
Suppose that
(a) $P(1)$ is true;
(b)   Whenever $P(k)$ is true then $P(k+1)$ is true also.
Then $P(n)$ is true for all natural numbers n.

In our present example $P(n)$ is the statement that
$S_n = n(n + 1)/2$.
We have already seen that condition (a) is satisfied.
Now suppose that for some natural number k, $P(k)$ is true.
Then

$$S_k = 1 + 2 + \ldots + k = k(k + 1)/2$$

So $S_{k+1} = k(k + 1)/2 + (k + 1)$

$$= (k + 1)((k/2) + 1)$$

$$= (k + 1)(k + 2)/2$$

But $P(k + 1)$ is that $S_{k+1} = (k + 1)((k+1) + 1)/2$, so
$P(k + 1)$ is true.

Hence condition (b) is also satisfied and the Principle
of Mathematical Induction then implies that $P(n)$ is true
for all n; that is:

$$1 + 2 + 3 + \ldots + n = n(n + 1)/2$$

for all natural numbers n, as was required.

We emphasise that in all mathematical induction
proofs it is necessary to prove:

(a)   that $P(1)$ is true
(b)   that if $P(k)$ is true for any k then $P(k + 1)$ is
      also true.

CHAPTER ONE

BASIC CONCEPTS OF DIGRAPHS

1.1   Relations

Mathematics is principally concerned with pattern, in
the widest sense of that term: not just geometrical
patterns, but in every situation or set of occurrences in
which come kind of regularity can be observed.   In physics
and some other branches of science the mathematics
appropriate to describing some of the most important
regularities has long been known.   In biological and
social sciences this mathematics is still in the making.
In other branches again the mathematics may not yet exist
or be recognised.   But in all fields of knowledge
mathematics is a way of describing regularities.

One common pattern is that objects of one kind are
related to objects of another kind in a consistent way.
For instance we may relate people to their homes, people
to books they have written, crops to the pests that eat
them.   Each member of one set may be related to a number
(perhaps zero) of members of the other.

The mathematical description of this situation is
called a (binary) relation.   We have two sets S and T,
and a relation which relates members of S with members of
T.   We can record which members are related with which by
means of a set of ordered pairs, the first component of
each being a member of S, and the second a member of T
related with it.   If $a \in S$ and $b \in T$ then (a,b) is a
member of this set when a is related to b.

For example consider the set of boys Allan, Billy,
Charlie, Don, and the set of girls Ethel, Florence, Gail.

Allan cares only for Ethel.   Billy and Charlie are
both keen on Florence, but Charlie is also interested in

Gail. Don is a confirmed bachelor and has nothing to do
with any of them.

Here S is the set of boys, T the set of girls, and the
relation is 'is fond of'. We can then record the
relationships by the ordered pairs (a,e), (b,f), (c,f),
(c,g), using people's initials to stand for their names.
There is no necessity for the sets S and T to be distinct.
Indeed there are very many examples in which they are
identical and this book is about relations between members
of the same set.

Sets of people yield many examples of such relations.
The family is a prolific source of relations such as 'x is
the father of y', 'p is the brother of q', 'u has the
same mother as v'. Here the letters represent people, and
each statement relates two people. But these two people
need not be distinct: in the last example, 'u has the
same mother as u' must always be true. In other examples
'x is related to x' may be true for some x and false for
others.

Turning to numerical examples, we consider the
relation on the positive integers, 'm divides n'. Thus '2
divides 6' is true, but '2 divides 5' is false. Every
number divides itself, so 'm divides m' is always true.
If we take the example 'm + 6 divides 6m', then 3 is
related to 3, and 6 to 6, but, for example, 4 is not
related to 4.

We need some terminology and some way of setting out
the information.

Each relation we are discussing is defined on some
set. This set will be called the domain of the relation.
The relation is a statement about pairs of members of this
domain. For each pair the statement is true or false.

Let D be the domain, and R the relation. Then we may
write 'a R b' to mean that 'a is related to b', and also
consider the corresponding set of ordered pairs (a,b).

If 'a R a' is true for all a in the domain we say that
R is reflexive. If 'a R a' is true for no a in D we say
that R is antireflexive. There are relations which are
neither reflexive nor antireflexive.

Besides listing the members of the domain and the
ordered pairs of the relation we can also represent a
relation by an 'arrow diagram'. We represent the members
of the domain by points (or, in practice, circles, squares
or other shapes) and draw a line from a to b if (a,b) is
an ordered pair of R. We put an arrowhead on the line to
show the direction from a to b. If (a,b) and (b,a) are
both true, then an arrow is drawn in each direction.

Suppose we have, for example, a set of people

$$\{a,e,i,o,u,b,c,d,f,g,h\},$$

in which case the vowels are assigned to females and the consonants to males. Suppose the relation 'is a parent of' on this set is described by the set of ordered pairs

$$\{(a,c),\ (a,i),\ (a,o),\ (b,c),\ (b,i),\ (b,o),\ (e,u),$$

$$(c,u),\ (o,f),\ (o,g),\ (o,h),\ (d,f),\ (d,g),\ (d,h)\}.$$

The arrow diagram in Fig. 1.1 shows this same information in a manner more easily comprehended. We have further distinguished the sexes by using circles for females and squares for males.

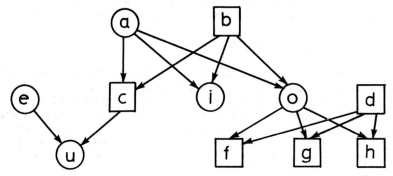

FIGURE 1.1

From this relation we may also define others: for the same set of people the relation 'has the same mother as' is shown in Figure 1.2.

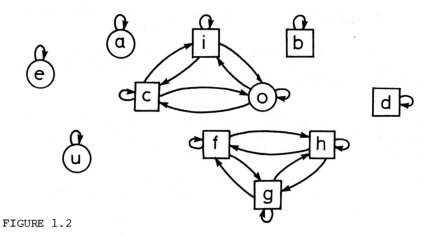

FIGURE 1.2

The student should write down the set of ordered pairs.
In this example every point has an arrow both beginning and
ending at the point. We call such an arrow a loop. In a
reflexive relation such as this, every point has a loop.
In an antireflexive relation no point has a loop.

In this book we are concerned mostly with antireflexive
relations on finite domains. Such a relation is called a
digraph. Thus the arrow diagram in Figure 1.1 represents
a digraph, but that in Figure 1.2 does not. We shall
define digraphs formally in section 1.3, and in the
succeeding chapters set up definitions and prove theorems
about digraphs. In many cases these definitions and
theorems can be extended to all relations on a set. This
is done, for instance, in [Anderson]*, where the term
relagraph is used for a relation treated in this way.

1.2 Equivalence Relations

Before settling to the main theme of the book we need
to discuss another kind of relation on a set.
Equivalence relations describe the concept of
'sameness'; that one member is the same as another, in some
sense. In Figure 1.2, for example, the relation is having
the same mother. The relation is an equivalence relation.
The properties defining an equivalence relation may be
expressed as follows:

Reflexive: Every member of the set is related to
               itself.

Symmetric: If a is related to b, then b is related
               to a.

Transitive: If a is related to b and b is related to
               c, then a is related to c.

For any equivalence relation we can define the
equivalence class of each member of the set to be the set
of objects related to that member:

$$[a] = \{x: x \text{ is related to } a\}.$$

In Figure 1.2 the equivalence class of c is $[c] = \{c,i,o\}$,
while $[a] = \{a\}$. We observe that two equivalence classes
either consist of exactly the same members or have no

* Footnote: Names appearing in parentheses are
             authors listed under 'Further reading'.

members in common. This feature holds for all equivalence
relations, and any relation for which it holds is an
equivalence relation.

If we list the equivalence classes, as

{a}, {b}, {c,i,o}, {d}, {e}, {f,g,h}, {u}

for the example in Figure 1.2, this collection of sets has
the properties that their union is the domain, and that the
intersection of any two distinct sets is empty. Any such
collection of sets is called a <u>partition</u> of the domain:
every partition of a set is the set of equivalence classes
for some equivalence relation on the set.

Anagrams of words give rise to another example of
equivalence relation. One word is an anagram of another
if it can be formed from it by rearranging the letters. In
order that each word shall be an anagram of itself we need
to define the null rearrangement as a rearrangement in
which the letters of the word are left in their original
places.

1.3 <u>Exercises</u>.

(S in the margin indicates that a solution or note on
the solution appears at the end of the book. * indicates
a harder question).

S 1. Let W = {stop, post, pots, sport, ports, spot, spots,
             top, tops, sort, port, pot, posts}.
Draw the arrow diagram and list the equivalence
classes of the anagram relation on this set.

S 2. With W as above, draw the arrow diagram of the
relation on W which relates words which differ in
exactly one place. Thus 'post' is related to 'port'
since they differ precisely in the third letter. Is
this relation an equivalence relation? Is it a
digraph?

3. With the same set again consider the relation in
which word X is related to Y if Y is obtained by
omitting any letter of X, keeping the other letters
in the same order. Thus 'sport' will be related to
'spot' and to 'port'. Is this an equivalence
relation? Is it a digraph?

4. Construct the ordered pairs and the arrow diagram for
the relation 'is a brother of' for the family shown
in Figure 1.1. Is it an equivalence relation? Is it
a digraph?

S 5. Heinrich (Evolution, 1975) reports the visiting of
seventeen types of flower by four species of bee
(A,B,C,D). The table below is selected from these
data.

	Flower	Bees
1.	Nymphaea	A,D,
2.	Scutellaria	C,D,
3.	Chelone	C,D,
4.	Salix	A,B,C,D,
5.	Rosa	A,B,D.

Construct the ordered pair representation of the
relation 'flower n is visited by bee X'. Construct
also the ordered pair representation and the arrow
diagram for the relation on the set of flowers 'every
species of bee which visits flower m also visits
flower n'.

6. Every function from a set to itself defines a
relation, in which f relates x to f(x).
Let $S = \{1,2,3,4,5,6,7\}$
Draw the arrow diagrams for the functions from S to S.
$f(x) = 8 - x$
$g(x) = x - 1$ if $x \neq 1$; $g(1) = 1$
$h(x) = x - 1$ if $x \neq 1$; $h(1) = 7$
$k(x) = \min \{x, 9 - x\}$
$p(1) = 2$, $p(2) = 6$, $p(3) = 5$, $p(4) = 2$, $p(5) = 6$,
        $p(6) = 1$, $p(7) = 7$.

1.4 Digraphs

DEFINITION 1.1 A digraph is a finite non-empty set
V, whose elements are called points, together with
a set A of ordered pairs (a,b), called arcs, where
a and b are distinct members of V.
We refer to a digraph by the ordered pair (V,A) or on
occasion by a single letter D.

DEFINITION 1.2 Two digraphs (V,A) and (W,B) are
equal if and only if V = W and A = B.

The terminology is by no means uniform: 'vertex' is
perhaps commoner than 'point', and 'node' is also used.
Instead of 'arc', 'directed edge' and 'branch' are used.
    The term 'digraph' is due to Harary, and is a
contraction of the words 'directed graph'. The longer term
is also in use. 'Network' is sometimes used in a general
sense: in Chapter 6 we give the term a more restricted

meaning. Although there are exceptions, we have generally
used the terminology of [Harary].
    The relationship between points and arcs is so
important that there are several ways of saying the same
thing.

DEFINITION 1.3 The following statements are exactly
equivalent.
(a) (a,b) is an arc;
(b) (a,b) is an arc from a to b;
(c) a is the beginning of the arc (a,b);
(d) b is the end of the arc (a,b);
(e) a is a precursor of b;
(f) b is a successor of a;
(g) (a,b) begins at a and ends at b.
Also we say that points a and b are adjacent if
either (a,b) or (b,a) is an arc.

    A digraph can be described by listing the points and
their successors. The successor table of the digraph in
Figure 1.1 is:

Point	Successors
a	c,i,o
e	u
i	
o	f,g,h
u	
b	c,i,o
c	u
d	f,g,h
f	
g	
h	

    This way of defining a digraph is more useful than the
ordered pair description in many of the techniques to be
described later. Correspondingly the digraph may be
described by listing the precursors of each point.

1.5 Indegree and Outdegree

    Looking at the arrow diagram representation of a
digraph, two of the most easily observed quantities are the
numbers of arcs ending or beginning at a point.

DEFINITION 1.4 The number of arcs beginning at the
point u is called the outdegree of u, and is
abbreviated od(u). The number of arcs ending at the

point u is called the <u>indegree</u> of u and is
abbreviated id(u).

Example.
Digraph A of Figure 1.3 has the following outdegrees
and indegrees:

Point	1	2	3	4	5	6
od	3	1	1	1	1	1
id	0	1	2	2	1	2

Observe that the sum of both the outdegrees and the
indegrees are 8, and that the digraph has 8 arcs.

DEFINITION 1.5 The following terms are applied to
points with certain indegrees and outdegrees.
A point u is a <u>source</u> if its indegree is zero
A point u is a <u>sink</u> if its outdegree is zero
A source is a <u>proper source</u> if its outdegree is non-
zero
A sink is a <u>proper sink</u> if its indegree is non-zero
A point is <u>isolated</u> if it is both a source and a sink;
that is, both indegree and outdegree are zero.

THEOREM 1.1   In any digraph the number of arcs is equal to
the sum of the outdegrees of the points, and also to the
sum of the indegrees.
     In formal notation:

$$\Sigma_{u \in V}\, od(u) \;=\; |A| \;=\; \Sigma_{u \in V}\, id(u)$$

PROOF. Each arc has exactly one beginning point. Each
point u is the beginning of exactly od(u) arcs. If we
add up the outdegrees we therefore count each arc exactly
once. The proof for the indegrees is exactly similar.   □
     The result further implies that the sum of the
indegrees and the sum of the outdegrees must be equal.
     A digraph D = (V,A) is termed <u>complete</u> if for every
pair of points u,v, both (u,v) and (v,u) are arcs of D.
That is:
     A = {(u,v): u ∈ V and v ∈ V, and u ≠ v}.

In a complete digraph there is an arc from each point
to each of the other points, so the outdegree (and
similarly the indegree) of each point is p - 1. Hence
summing over all points, there are
     p.(p - 1) arcs.
     Since every other digraph omits some of the arcs in
the complete digraph, no digraph can have more than

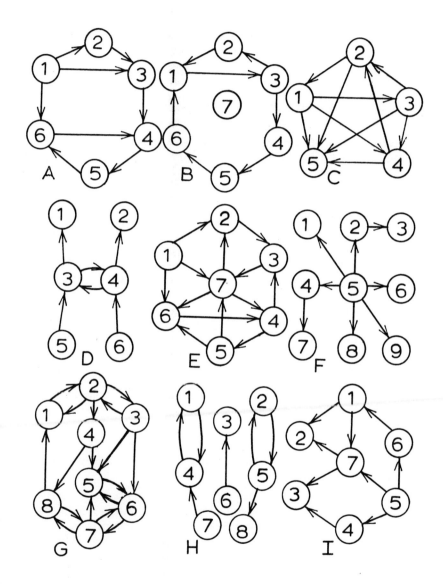

FIGURE 1.3

p.(p - 1) arcs, where p is the number of points.

1.6 Exercises

S 1. Write down the point set and the arc set of the
     digraph A in Figure 1.3.

S 2. Write down the point set and arc set of the digraph
     E of Figure 1.3.

S 3. List the precursors and successors, and indegree and
     outdegree for each point in digraph E of Figure 1.3.

* 4. Two points of a digraph are 'precursor-successor
     equivalent' if they have the same precursors and the
     same successors. Show that precursor-successor
     equivalence is an equivalence relation. Find the
     equivalence classes in Figure 1.1.

S 5. A child's model railway is made up of sections of
     track of five kinds, as shown in Figure 1.4. The peg
     at the end of one section fits into the hole on
     another section to link them together. There are many
     sections of each kind. A complete track is obtained
     by linking sections, so that no peg or hole is left
     exposed.
        Each completed track can be represented by a digraph,
     whose points represent the sections and in which arc
     (a,b) means that the peg on section a is fitted into
     the hole in section b.
        In such a digraph the indegree and outdegree of each
     point indicates the kind of section it represents.
        Use Theorem 1.1 to obtain an equation linking the
     numbers of ramps, buffers and branches.

  6. Draw arrow diagrams, list points and arcs, and
     construct successor tables for complete digraphs with
     1,2,3,4,5 vertices.

  7. On a certain island the only living things are
     cabbages, thistles, snails, thrushes, goats, lions and
     men. The snails eat cabbages, the thrushes eat snails,
     the goats eat cabbages and thistles, the men eat goats
     and cabbages and the lion eats goats and men.
        Construct a digraph whose points are the forms of
     life and whose arcs (a,b) mean that form a eats form
     b. Such a digraph is called a food web.

  8. Some flowers, such as Oxalis, have different
     arrangements of stamens and pistil on different
     plants. The pistil may be long, medium or short,

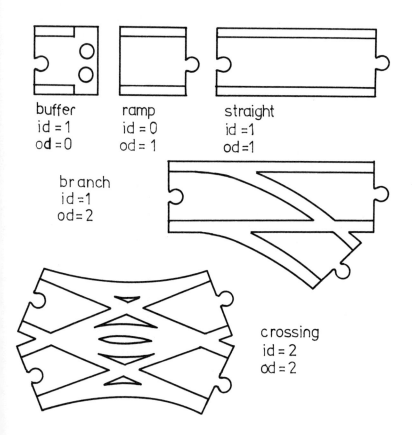

buffer
id = 1
od = 0

ramp
id = 0
od = 1

straight
id =1
od =1

branch
id =1
od= 2

crossing
id = 2
od = 2

FIGURE 1.4

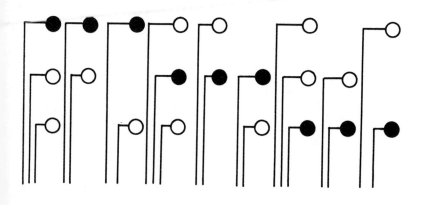

FIGURE 1.5

the stamens are then of the other two lengths. Pollen
can be transferred only from a stamen of one length to
a pistil of the same length (and therefore on a
different plant). If also it is possible for plants to
lack stamens of one length, there are nine possible
structures a - i in Figure 1.5 .
   Filled circles represent pistils and open circles
stamens.
   Construct a digraph whose points are a to i and in
which arc (x,y) means that pollen can be transferred
from a stamen of x to the pistil of y.

1.7 Walks

The concept which is most important in the theory of
digraphs and makes it more than just a branch of relation
theory is the walk.

> DEFINITION 1.6   A walk W in a digraph D = (V,A) is a
> finite sequence, consisting of points and arcs of D
> alternately, and beginning and ending with points.

$$W = u_1, \ (u_1, \ u_2), \ u_2, \ (u_2, \ u_3), \ \ldots, \ (u_{n-1}, u_n), \ u_n;$$

Each arc begins at the point written on its left and ends at
at the point written on its right. In W above, u_1 is
called its first point and u_n its last point. We shall
also say that W is a walk from u_1 to u_n.

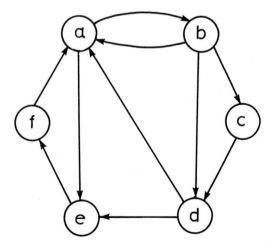

FIGURE 1.6

In Figure 1.6 there are walks

W = a,(a,b),b,(b,c),c,(c,d),d;

X = a,(a,b),b,(b,a),a,(a,e),e,(e,f),f,(f,a),a,(a,b),b.

This second example emphasises that a point or an arc
may occur more than once in a walk.

Although we shall always consider a walk to consist of
both points and arcs, it is unnecessarily cumbersome to
write walks in this manner. We can instead write just the
points or just the arcs and deduce the arcs or points
respectively from the fact that we are writing a walk. In
terms of arcs we can write W and X as:

W = (a,b),(b,c),(c,d)

X = (a,b),(b,a),(a,e),(e,f),(f,a),(a,b).

Even shorter is the expression in terms of points only:

W = $\langle$a,b,c,d$\rangle$

X = $\langle$a,b,a,e,f,a,b$\rangle$

In writing a walk in this form we shall always enclose
it in these pointed brackets $\langle,\rangle$, and we shall generally
use this most concise form.

When considering distance properties, and also in
Chapter 2 when we are studying the structure of digraphs ,
we find it advantageous to accept walks which have no arcs,
but consist of a single point; $\langle$u$\rangle$. One disadvantage of
the arcs-only notation is that it is impossible to write
such a walk at all.

We must define terms to describe various special
kinds of walk.

DEFINITION 1.7   A walk $\langle$u$\rangle$, which may be defined at
any point u, is called trivial.
A walk whose first and last points coincide is called
closed.
A walk all of whose points are distinct (and in which
therefore no arc is repeated either) is called a path.
A closed walk all of whose points are distinct
except the first and last is called a cycle.

The trivial walks are also described in appropriate
contexts as trivial paths and trivial cycles. Clearly no
other walk can be both a path and a cycle.

It should be noted that some writers use 'path' and
'cycle' in the senses in which we have used 'walk' and
'closed walk' respectively.

It is useful to associate with each walk a <u>length</u>, which is the number of arcs it contains, each counted as often as it occurs. Thus the walks W and X in the digraph of Figure 1.6 have lengths 3 and 6 respectively. In the points-only notation the length of a walk is equal to the number of commas. The trivial walks have length zero. Whenever there is a walk from u to v we can define a distance from u to v.

> DEFINITION 1.8 If there exists a walk from point u
> to point v of a digraph, then the <u>distance</u> from u to
> v is the minimum length of walk from u to v. We
> write the distance as d(u,v).
> If there is no walk from u to v we do not define the
> distance.

In the digraph of Figure 1.6 it is possible to find a walk from each point to each other point. We can therefore define d(u,v) for all u and v and set up a table as below.

	a	b	c	d	e	f
a	0	1	2	2	1	2
b	1	0	1	1	2	3
c	2	3	0	1	2	3
d	1	2	3	0	1	2
e	2	3	4	4	0	1
f	1	2	3	3	2	0

The distance d(u,u) is always defined as zero. We may observe that d(u,v) and d(v,u) are not always equal: indeed one may be defined and not the other.

> DEFINITION 1.9 A walk (cycle) is a <u>spanning</u> walk (cycle)
> if it contains all the points of the digraph.

The digraphs we will define later as <u>strong</u> (Section 2.9) are precisely those with a spanning closed walk.
In the digraph of Figure 1.5.1 ⟨b,c,d,a,e,f⟩ is a spanning path and ⟨c,d,e,f,a,b,d⟩ is a spanning cycle.

1.8 Joining and splitting walks

Consider the two walks
$$W = ⟨a,b,c,d⟩$$
$$Z = ⟨d,a,e⟩$$
in the digraph of Figure 1.6. The last point of W is the same as the first point of Z. We can then <u>join</u> W and Z

to make a new walk $\langle a,b,c,d,a,e \rangle$ which we will call WZ.

We can join two walks only if the last point of the walk on the left coincides with the first point of the walk on the right. Therefore to write ZW for these two walks would be meaningless. If P and Q are two walks then PQ and QP are both defined if and only if the last point of each is the first point of the other. If this happens then QP and PQ are both closed walks. For instance in Figure 1.6 there are the walks P = $\langle b,c,d \rangle$ and Q = $\langle d,a,b \rangle$. Then PQ = $\langle b,c,d,a,b \rangle$ and QP = $\langle d,a,b,c,d \rangle$.

These two closed walks contain the same points and the same arcs in the same cyclic order, differing only in their 'choice' of first and last points. This is a distinction we shall sometimes want to ignore.

Just as we have joined walks, so we may split a walk. That is, we may express it as the join of two or more walks. The walk X = $\langle a,b,a,e,f,a,b \rangle$ in Figure 1.6 may, for instance, be expressed as X = TUT, where T = $\langle a,b \rangle$ and U = $\langle b,a,e,f,a \rangle$.

Splitting and joining help us to prove the following results, of which the first is perhaps the most important in the whole theory of digraphs.

THEOREM 1.2   If u,v,w are points in a digraph such that there is a walk from u to v and a walk from v to w, then there is a walk from u to w.

PROOF. Let P be a walk from u to v, and Q a walk from v to w.   Then PQ is a walk from u to w.   □

THEOREM 1.3   If u and v are points in a digraph such that there is a walk from u to v and a walk from v to u, then there is a closed walk whose first and last point is u and which passes through v.

PROOF. Put w = u in the previous theorem.   □

THEOREM 1.4   If u and v are points in a digraph such that there is a cycle whose first and last point is u, and which passes through v, then there is a path from u to v and a path from v to u in the digraph.

PROOF. The cycle can certainly be written as a join of walks P from u to v and Q from v to u. If either P or Q is not a path then some point occurs twice in one of them. But then PQ would not be a cycle. Hence P and Q are paths.   □

Indeed the paths P and Q have only their first and last points in common. If we are given that there is a

path from u to v and a path from v to u this does imply a
closed walk whose first and last point is u and which
passes through v, but not that it is a cycle, or that any
such cycle exists.

THEOREM 1.5   If P and Q are walks such that the last point
of P is the first point of Q, then
$$\text{length } (PQ) = \text{length } (P) + \text{length } (Q)$$

PROOF. The length of each walk is the number of arcs,
counted as often as they occur.   □

THEOREM 1.6   (Triangle inequality). Let u,v,w be points
in a digraph such that $d(u,v)$ and $d(v,w)$ are both defined.
Then so is $d(u,w)$ and
$$d(u,w) \leqslant d(u,v) + d(v,w).$$

PROOF. If $d(u,v)$ is defined there is a walk P whose first
point is u and last point v, and whose length is $d(u,v)$.
If $d(v,w)$ is defined there is a walk Q whose first point
is v and last point w, and whose length is $d(v,w)$.
Then by THEOREM 1.2, PQ is a walk from u to w, so $d(u,w)$ is
defined. Further,
$$\text{length } (PQ) = \text{length } (P) + \text{length } (Q)$$
$$= d(u,v) + d(v,w)$$
As PQ may not be the shortest walk from u to w we know only
that $d(u,w) \leqslant \text{length } (PQ)$
     Hence $d(u,w) \leqslant d(u,v) + d(v,w)$ □

It is important to notice that these results hold even
if two, or all three points coincide, in which cases some
of the walks may be trivial.
[We will examine distances in more detail when we come
to discuss strong digraphs in section 2.9, and some
considerations will be deferred until we discuss digraphs
in the language of matrices in Chapter 8].
If a walk is not a path, then some point occurs more
than once. Intuitively we may then omit the section of the
walk between two occurrences and obtain a shorter walk, with
the same first and last points. We can represent any walk
W which is not a path as the join of three walks, W = XYZ,
in which Y is a nontrivial closed walk. Then XZ is a
shorter walk with the same first and last points as W. If
XZ is not a path we can extract another closed walk. As
the walk remaining is getting shorter at each stage, the
process must stop eventually, and when this happens we must
have a path. Thus if there is a walk from any point u to
a point v, then there is also a path from u to v.

If W = $\langle$t,h,i,s,w,a,l,k,i,s,a,n,e,x,a,m,p,l,e$\rangle$,
    Y = $\langle$a,l,k,i,s,a$\rangle$,
    Z = $\langle$a,n,e,x,a,m,p,l,e$\rangle$,
the walk XZ extracted is
        $\langle$t,h,i,s,w,a,n,e,x,a,m,p,l,e$\rangle$
This is still not a path as a and e are repeated. We can
extract walks $\langle$t,h,i,s,w,a,m,p,l,e$\rangle$ and $\langle$t,h,i,s,w,a,n,e$\rangle$,
which are both paths. We thus observe that not only may we
take different reduction steps, but there is no uniqueness
of end result.

THEOREM 1.7   If W is a walk from u to v and
length (W) = d(u,v), then W is a path.

PROOF. If W is not a path, then by the above process we
can find another walk with the same first and last points,
but shorter. But no walk can be shorter than the distance
between its first and last points. Hence W must be a path.
$\square$

But this theorem does not assert that every path from
u to v has length d(u,v), and that statement is false, as
the example of the process shows, since we have found two
paths with different lengths.

In the same way that we extracted a path from a walk,
we can extract a non-trivial cycle with the same first and
last point from any non-trivial closed walk. The proof
does not follow directly from the process above, but a
parallel argument may be devised.

1.9 An Application to logic

We are now in a position to undertake an application
of digraphs: to the solution of a certain class of
problems in logic. Consider the following set of
propositions propounded by C.L. Dodgson (Lewis Carroll),
one of the founders of symbolic logic.

(i)   No kitten that loves fish is unteachable.
(ii)   No kitten without a tail will play with a
       gorilla.
(iii) Kittens with whiskers always love fish.
(iv)   No teachable kitten has green eyes.
(v)   No kittens have tails unless they have whiskers.

First we must study the structure of these
propositions. Each links two of the following statements
about kittens, each of which may be true or false about a
particular kitten k:

L: k loves fish   E: k has green eyes
T: k is teachable   P: k will play with a gorilla
W: k has whiskers   C. k has a tail

To each such statement corresponds its <u>negation</u>;   for example

        not L: k does not love fish
        not C: k does not have a tail

The five propositions can then be written   .

        (i)   If L then T
        (ii)   If not C then not P
        (iii)   If W then L
        (iv)   If T then not E
        (v)   If C then W

The original forms by no means exhaust the alternative
ways the English language has of expressing (or disguising)
"If X then Y". The situation is not made easier that each
statement has a <u>contrapositive</u> "if not Y then not X" which
has the same meaning;   that is, either a statement and its
contrapositive are both true or they are both false, and
either may equally be used to represent in symbolic form
the original verbal statement.   In our approach we avoid
having to choose between them by representing each
proposition by the two symbolic propositions together.
Remembering that "not (not X)" is the same as "X", the
contrapositives of Carroll's propositions are:

        (ic)   If not T then not L
        (iic)   If P then C
        (iiic)   If not L then not W
        (ivc)   If E then not T
        (vc)   If not W then not C

From these propositions Carroll deduces that "no kitten
with green eyes will play with a gorilla", that is, the
pair of symbolic propositions

        "if E then not P" and "if P then not E"

The key to Carroll's approach, and our own, is to use
the logical principle that if (if X then Y) and (if Y then
Z), then (if X then Z).   Our approach is to exploit the
identical structure of Theorem 1.2; if there is a walk from
x to y, and a walk from y to z, then there is a walk from
x to z.   We therefore construct a digraph whose points are
the basic statements and their negations, in this case,
L,T,W,E,P,C, not L, not T, not W, not E, not P, not C.   We
wish to represent each statement "if X then Y" by a walk:
we do this by making the arcs of the digraph the
corresponding arcs (X,Y), and so creating walks $\langle X,Y \rangle$.
Every walk in the digraph then represents a valid
deduction from the given propositions.   Because each

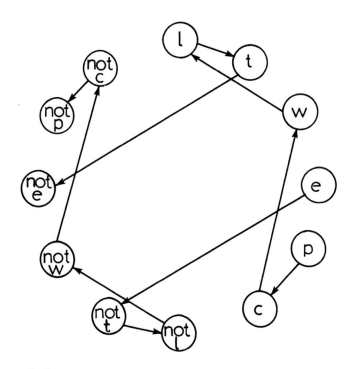

FIGURE 1.7

proposition also has its contrapositive in the digraph,
each proposition we deduce will also have its contrapositive
in the digraph.

     Figure 1.7 shows the digraph. As there is a walk from
P to not E, and a walk from E to not P, we may deduce, as
required, that "if E then not P", "no kitten with green
eyes will play with a gorilla".

     The technique for this kind of problem is thus to list
the basic statements and translate the verbal propositions
into pairs of symbolic propositions in terms of the basic
statements and their negations. The statements and their
negations become the points of a digraph, and arcs (X,Y),
(not Y, not X) correspond to the pair of propositions 'if
X then Y' and 'if not Y then not X'. The consequences of
the propositions may then be read as walks in the digraph:
any proposed proposition which does not correspond to a
walk is not a consequence of the given propositions.

     We may get a digraph in which there is a walk from X
to not X. This shows   that the propositions imply that X
must be false. If there is not only a walk from X to not X
but also a walk from not X to X, then the system of given

propositions is inconsistent. If it should happen that
there is a closed walk which does not include any statement
and its negation then the statements represented by points
through which the walk passes are all true or all false
together.

1.10 Exercises

S 1. A digraph has a walk
$$Y = \langle x,e,f,h,x,y,e \rangle.$$
Write this walk in full (point-and-arc) notation and
in arc-only notation. Is Y a path? Is Y a closed
walk?

S 2. Find which of the digraphs in Figure 1.3 have (a) a
spanning walk, (b) a spanning path, (c) a spanning
closed walk, (d) a spanning cycle.

S 3. Show that every digraph with a spanning cycle has a
spanning path.

4. Construct a digraph whose points are W (for walk),
P (for path), T (for trivial walk), K (for closed
walk), C (for cycle) and SW, SP, SK, SC for their
spanning counterparts. The arcs (X,Y) of the digraph
are to indicate that every X is a Y. (e.g. (P,W) is
an arc since every path is a walk).

5. Construct the distance table for digraph F of
figure 1.3.

S 6. Suppose D is a digraph with p points with a spanning
path. What is the length of a spanning path of D?

7. Figure 1.8 represents part of the road map of
Christchurch, New Zealand as it was in 1976. Points
represent road intersections and arcs legal
directions of vehicle travel. PB is a parking
building and BR is the Bridge of Remembrance. Find
a cycle of minimum length which has the parking
building as first and last point and also includes the
Bridge of Remembrance.

8. In Rugby Union Football the score of each team can
advance in steps of 3,4,6 points. Construct the
digraph whose points are the integers from 0 to 17,
and whose arcs are all pairs of the forms (n, n + 3),
(n, n + 4), (n, n + 6) with values in this range.
Which numbers are not possible points totals? For
each possible total find the smallest number of steps
to reach it from 0.

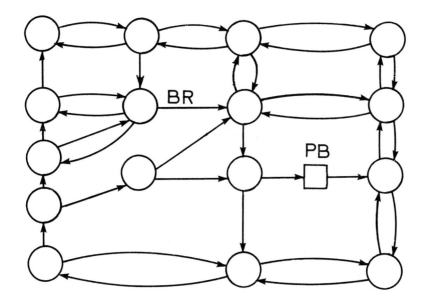

FIGURE 1.8

9. Let $\langle a,c,d,b,e,f,a,b,c,f,g,h,e\rangle$ be a walk in a digraph. Extract all the possible paths from a to e contained in this walk. Construct the digraph whose points and arcs are precisely those implied by the above walk, and show that there is a path from a to e in this digraph shorter than any that can be extracted from the given walk.

S 10. The five most important forms of energy are thermal, mechanical, electrical, chemical and radiant. Representing these by the points of a digraph, and using initials for brevity, the most important interchanges are:
(t,m),(m,t),(m,e),(e,m),(e,t),(e,c),(e,r),(c,t),(c,e), (r,t),(r,c).
Draw the digraph and construct its distance table. Find a spanning cycle of the digraph.

11. Show that if there is a closed walk through two points u and v, then the length of the shortest closed walk through them both is d(u,v) + d(v,u). Show that even if there is a cycle through u and v there may not be

any cycle of length d(u,v) + d(v,u).

S 12.We have not made any definition of equality of two
    walks, although such a definition is implicit in all
    equations involving walks. Formulate a suitable
    definition.

S 13.Given the statements A,B,C,D and their negations and
    the propositions 'If A then not D', 'If not B then D',
    'If B then not C', 'If D then not C', which of the
    following can be proved and which not? 'If C then
    not A', 'If A then D', 'not C', 'not B'.

  14.(Lewis Carroll) What can be deduced from:
          All well-fed canaries sing loud;
          No canary is melancholy if it sings loud.

  15.(Lewis Carroll) What can be deduced from:
          No Frenchmen like plum pudding;
          All Englishmen like plum pudding.

  16.(Lewis Carroll) What can be deduced from:
          Babies are illogical;
          Nobody is despised who can manage a crocodile;
          Illogical persons are despised.

  17.Draw the digraph corresponding to:
          Jubjubs are birds;
          All birds can fly;
          No jubjubs can fly.
    Comment on the conclusion.

1.11 Subdigraphs and partial digraphs

    Figure 1.9 shows the 'Carbon Cycle', the various
states in which carbon exists in the world and the ways in
which it is transferred from one state to another. This
digraph is of course much simplified.
    We might however be interested in only part of this
system. We could restrict our attention in either of two
ways. We could consider just some of the points and all
arcs between them, or all the points and some of the arcs
between them. We could also consider some of the points
and some of the arcs, provided that for each arc
considered we take both its beginning and end point in the
set of points. Such a restriction of attention can be
considered as a combination of the other two restrictions.

    DEFINITION 1.10 If D = (V,A) is a digraph, and U
is a non-empty subset of V, then the digraph

(U, A ∩ (U×U)), whose point set is U and whose arcs
are those arcs of D which both begin and end in U, is
termed the subdigraph of D defined by U.

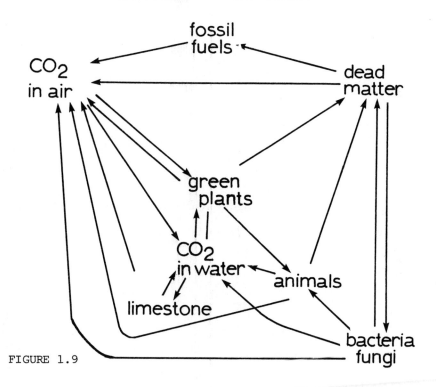

FIGURE 1.9

As an example, if all life suddenly vanished from the
earth only the inanimate states and inanimate processes
between them would remain, as shown in Figure 1.10.

DEFINITION 1.11 If D = (V,A) is a digraph and
B ⊆ A, then the digraph (V,B) is termed a partial
digraph of D.
As an example, we start from the carbon cycle again,
and consider only those arcs representing transfers by
respiration and photosynthesis. We include all the points.
The digraph is shown in Figure 1.11.
A digraph formed by a combination of these methods is
called a partial subdigraph of the original digraph.

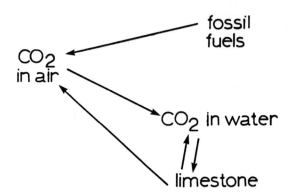

FIGURE 1.10

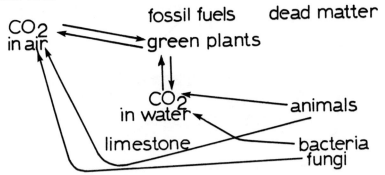

FIGURE 1.11

1.12 Converse

In Figure 1.12 we show two digraphs D and E with the same point set.

Notice that where D has an arc (a,b) then E has the arc (b,a) and vice-versa. We call D and E converse of each other. Note that arcs (4,5) and (5,4) are present in both D and E.

DEFINITION 1.12   If A is any set of ordered pairs, we define A^R, the reverse of A, by

$$A^R = \{(b,a): (a,b) \in A\}.$$

Observe that the reverse has the following properties:

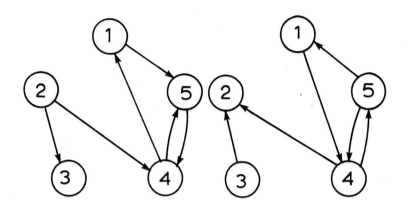

FIGURE 1.12

THEOREM 1.8   If A and B are sets of ordered pairs then

    (a) $(A^R)^R = A$

    (b) $A^R \subseteq B^R$ if and only if $A \subseteq B$

    (c) $A^R \cup B^R = (A \cup B)^R$

    (d) $A^R \cap B^R = (A \cap B)^R$

The PROOF is left to the reader.

    DEFINITION 1.13 If $D = (V,A)$ is any digraph, we define its converse D^C to be the digraph (V,A^R).
    Part (a) of Theorem 1.8 implies at once that $(D^C)^C = D$.

THEOREM 1.9 If D is any digraph and there is a walk from a to b in D, then there is a walk from b to a of the same length in D^C.

PROOF Outline: This walk consists of the same points and arcs as the given walk, but with each arc reversed and the order of the whole reversed.   □
In some digraphs there may be arcs in both directions between some pairs of points. In one important class of digraphs, whenever (u,v) is an arc, (v,u) is also an arc.

    DEFINITION 1.14 A digraph (V,A) is <u>symmetric</u> if whenever (u,v) ∈ A, then (v,u) ∈ A also.
    This meaning of 'symmetric' is exactly the same as that for a symmetric relation in section 1.2.
    Examples include such relations between people as

'has shaken hands with', 'is married to'. Geographical
examples include 'shares a common boundary with' as applied
to regions.
        Many properties take on specially simplified forms for
symmetric digraphs. It is usual to replace the pair of
arcs (u,v), (v,u) by a <u>line</u>, the unordered pair {u,v}. We
will pursue the properties of the resulting <u>graph</u> in
Chapter 9.

<u>THEOREM 1.10</u> A digraph is symmetric if and only if it is
equal to its converse.
The PROOF is left to the reader.

<u>THEOREM 1.11</u> A digraph is symmetric if and only if d(v,u)
is defined for precisely the same pairs of points as
d(u,v), and also that
$$d(v,u) = d(u,v)$$
whenever either is defined.
PROOF. Suppose first that the digraph is symmetric.
Suppose that
        $<u, (u,v), w, (w,x) x, ..., (z,v), v>$
is a walk with first point u and last point v. Then the
digraph must contain arcs (u,w), (w,x), ..., (z,v). Being
symmetric, the digraph must also have arcs (v,z), ..., (x,w),
(w,u), so
        $<v, (v,z), z, ..., x, (x,w), w, (w,u), u>$
is a walk, of the same length from v to u. Hence if d(u,v)
is defined, so is d(v,u), and if the given walk is of
minimal length, d(u,v), then the reversed walk demonstrates
that d(v,u) is not greater than d(u,v). Since any walk from
v to u can similarly be reversed, the two distances must be
equal.
        Conversely, suppose that d(u,v) = d(v,u) where either
is defined. We consider only the pairs of points where the
distance is 1. The arc set of the digraph is
        $A = \{(u,v) : d(u,v) = 1\}$
        But if d(u,v) = 1 then d(v,u) = 1 also. Hence if
(u,v) ∈ A then (v,u) ∈ A. So the digraph is symmetric.   □

1.13 Isomorphism

        Figure 1.13 shows $D_1 = (V_1, A_1)$, a digraph whose
points represent vegetables and whose arcs represent the
assertion that the plant a grows better when planted close
to plant b.
        Figure 1.14 shows $D_2 = (V_2, A_2)$, a digraph whose
points represent positive integers and whose arcs (a,b)

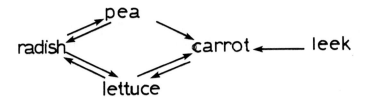

FIGURE 1.13

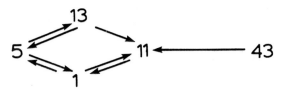

FIGURE 1.14

indicate that 2a + b and 3a + 4b are both prime.

These digraphs deal with very different applications, yet as digraphs they correspond very closely. If we construct a table:

V_1	pea	carrot	lettuce	radish	leek
V_2	13	11	1	5	43

and we replace in D_1 the names of the vegetables by the numbers corresponding to them in the table, we transform Figure 1.13 into Figure 1.14. Similarly if we replace the numbers in Figure 1.14 by the vegetable names corresponding to them, we obtain Figure 1.13. We call two digraphs which differ only in the names assigned to the points isomorphic to each other.

Formally we interpret the table as defining an invertible function f mapping V_1 onto V_2 so that f(pea) = 13 etc. Reading the table upwards gives the inverse function f^{-1} mapping V_2 onto V_1. The special property of this function, as distinct from all other invertible functions from V_1 onto V_2 is that it 'preserves' arcs. That is, if (a,b) is an arc in D_1, then (f(a),f(b)) is an arc in D_2, and all the arcs of D_2 can be obtained from arcs in D_1 in this manner.

DEFINITION 1.15 Let $D_1 = (V_1,A_1)$ and $D_2 = (V_2,A_2)$ be two digraphs. An invertible function f mapping V_1 onto V_2 is an isomorphism if:

$$A_2 = \{(f(u),f(v)): (u,v) \in A_1\}.$$

DEFINITION 1.16   If there exists an isomorphism from D_1 to D_2. we say that D_1 is _isomorphic_ to D_2.

THEOREM 1.12   'Being isomorphic' is an equivalence relation.

EXPANSION OF THEOREM.   As set out in Section 1.2, three properties must be verified to establish an equivalence relation.   These are:

(i)   Every digraph is isomorphic to itself.

(ii)   If digraph D_1 is isomorphic to D_2, then D_2 is isomorphic to D_1.

(iii) If D_1 is isomorphic to D_2, and D_2 is isomorphic to D_3, then D_1 is isomorphic to D_3.

PROOF OUTLINE.

(i)   The identity function on V_1, $i(u) = u$, is an isomorphism from D_1 onto itself.

(ii)   If f is an isomorphism from D_1 to D_2, then f^{-1} is an isomorphism from D_2 to D_1.

(iii) If f is an isomorphism from D_1 to D_2, and g is an isomorphism from D_2 to D_3, then gof is an isomorphism from D_1 to D_3.   □

Two digraphs may now be described as isomorphic to each other. Isomorphism is important because isomorphic digraphs differ only in the labelling of the points. They coincide in all purely digraph properties. Thus two isomorphic digraphs must have the same number of points and the same number of arcs. They have the same number of points of each indegree and each outdegree. Walks must also coincide. These and other properties are examined in the problems below.

These results are easy to use in a negative way, for failure of two digraphs to agree over any one of these isomorphism invariants proves them to be non-isomorphic. Except in very rare circumstances, that two digraphs are isomorphic is proved only by producing an isomorphism.

Where an isomorphism exists it may not be unique. In particular there may exist non-identity isomorphisms from a digraph to itself. These are called automorphisms of the digraph and form a group under composition. While this

group is important in the abstract study of digraphs, it
will not be discussed further in this book. Chapters on
groups and graphs are included in [Anderson], [Harary].

1.14 Exercises

1. Prove that if a digraph D has a partial digraph H which
    has a spanning walk, then D also has a spanning walk.
    If D has a subdigraph which has a spanning walk, does
    that prove that D has a spanning walk?
2. Construct from Figure 1.1 two partial digraphs whose
    arcs show respectively the relations 'is the father of'
    and 'is the mother of'.
3. Let (V,A) be any digraph. Prove that $(V,A \cap A^R)$ and
    $(V,A \cup A^R)$ are both symmetric. Find these digraphs for
    G of Figure 1.3.
4. Show that every subdigraph of a symmetric digraph is
    symmetric. Is every partial digraph of a symmetric
    digraph symmetric?
5. Prove Theorem 1.8.
6. Prove Theorem 1.10.
7. Figure 1.15 shows the directions of traffic in the
    major streets of an English country town. Points a to
    f indicate routes leading to other towns. At each

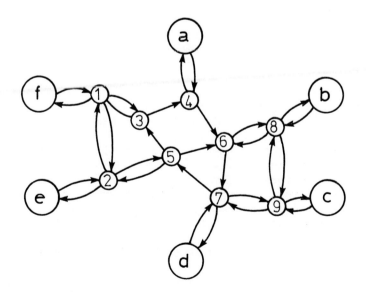

FIGURE 1.15

7. (continued)
    intersection a road sign must be erected directing
    through traffic to each destination. The designated
    arcs from each point define a partial digraph for each
    destination. Construct a suitable partial digraph for
    each destination.

S 8. The routes mentioned in Question 7 above would avoid
    the central one-way streets as much as possible. Call
    a section of one-way street <u>avoidable</u> if there is a walk
    from its beginning to its end consisting only of two-way
    streets. Determine for each one-way street in
    Figure 1.15 whether it is avoidable or not. Can you
    devise a general criterion which is easy to operate?

*9. Prove that if f is an isomorphism from D_1 to D_2 which
    maps point u of D_1 to point v of D_2, then
    $id(u) = id(v)$ and $od(u) = od(v)$.

S 10. Find two different isomorphisms between the digraphs
    (V,A) and (W,B), where
    $V = \{1,2,3,4,5\}$ $A = \{(1,2),(2,3),(3,4),(4,1),(5,2),$
    $(5,4)\}$.
    $W = \{a,b,c,d,e\}$ $B = \{(a,d),(a,e),(c,e),(d,c),(b,d),$
    $(e,b)\}$.

S 11. If $X = \{p,q,r,s,t\}$ and
    $C = \{(p,r),(r,s),(s,t),(t,p),(q,p),(q,r)\}$,
    prove that (X,C) is not isomorphic to (V,A) of
    question 10.

12. Prove that if there is an isomorphism f from D_1 to D_2,
    and if there is a walk from u to v in D_1, then
    there is a walk of the same length from f(u) to f(v) in
    D_2. Show that if the first walk is a path, then so is
    the second.

13. Show that if a digraph D is isomorphic to a digraph E,
    and E has a spanning cycle, then so has D.

*14. Prove that if a digraph D is isomorphic to digraph E,
    then D^c is also isomorphic to E^c.

15. A digraph is called <u>self-converse</u> if it is isomorphic
    to its converse. Show that digraph D of Figure 1.3
    is self-converse. Show that F on the same page is not
    self-converse. Test each of the other digraphs on the
    page for self-converse property.

*16. Show that every logic digraph, as developed in 1.7 is
    self-converse.

*17. Show that every digraph isomorphic to a self-converse
    digraph is self-converse.

18. Show that every digraph isomorphic to a symmetric
    digraph is symmetric.

CHAPTER TWO

DIGRAPH STRUCTURE

2.1 Structure

The study of structure is one of the main occupations
of the mathematician. In an application of mathematics,
the mathematical structure can often be used to illuminate
the problem. In the field of digraphs the process of
analysing the structure of a digraph is analogous to the
dissection of an animal or plant. The biologist will be
concerned with isolating and identifying the functions of
the organs of the creature and also with studying their
interrelations. In the same way we shall be concerned with
identifying certain subdigraphs, the components,
corresponding to the organs, and to construct a new digraph,
the condensation, which describes the interrelation between
the components.

Just as the biologist will classify the creature on
the basis of the organs and their interrelations, so we will
be able to classify digraphs according to the nature of
their components and condensations. In various applications
too, the components and condensation can help us understand
the system being described.

Another benefit of this analysis is to be able to draw
the arrow diagram of a digraph in such a way that it is
most easily understood, in the same way that a biologist
might display a specimen. Consider the arrow diagram in
Figure 2.1. It is a mess. There is no discernable
pattern. The analysis carried out in the next three
sections enables us to redraw the diagram as in Figure 2.5
in such a way that the basic structure can be taken in at
a glance.

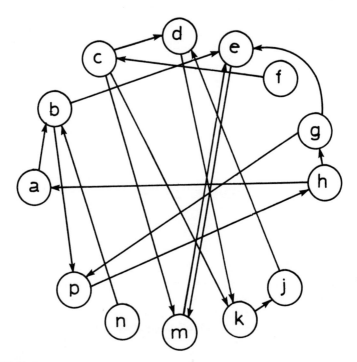

FIGURE 2.1   :

2.2   Reachability and Equivalence

In the elucidation of digraph structure the
fundamental tool is the walk. But the details of a
particular walk from u to v are unimportant; we are
interested only in whether a walk exists with u as its
first point and v as its last point. We therefore recast
the statements of sections 1.7 and 1.8 to remove the
references to specific walks.

DEFINITION 2.1 If u and v are points in a digraph D,
and there is a walk with first point u and last
point v, then we say that v is <u>reachable</u> from u.

We will find it convenient to work mostly with sets
of points so we define:

DEFINITION 2.2 If u is a point in a digraph D we
define the <u>inreach</u> $R^I(u)$ and the <u>outreach</u> $R^O(u)$ to be
the sets:

$$R^I(u) = v: \{u \text{ is reachable from } v\}$$
$$R^O(u) = w: \{w \text{ is reachable from } u\}$$

For example, part of the schedule of courses in a university mathematics department might read:

Course	Prerequisite
LA1	-
NS1	-
C1	-
LA2	LA1 and NS1
AA2	LA1 and NS1
MP2	LA1 and C1
LA3	LA2
AA3	AA2
MP3	LA2 and MP2

(LA-linear algebra; NS-number systems; AA-abstract algebra; C-calculus; MP-mathematical programming).

In order to enrol for each course each prerequisite course must be passed. If we interpret the prerequisite table as a precursor table we obtain a digraph as in Figure 2.2.

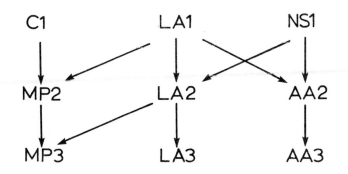

FIGURE 2.2

The inreach of each course consists of the course itself and all those courses which must be passed before the given course can be taken. For example,

$$R^I(MP3) = \{MP3, MP2, LA2, C1, LA1, NS1\}$$

and the outreach of each course consists of the course

itself and all the courses for which, directly or
indirectly, it is required preparation.

$$R^O(C1) = \{C1, MP2, MP3\}$$

In every digraph each point belongs to both its inreach
and its outreach, because of the trivial walks. In this
particular application we would be as happy if the point
were excluded from these sets, but for the purposes of
investigating structure, inclusion of the point yields much
simpler results.

As a further example consider the inreach and outreach
sets for the digraph H of Figure 1.3.

u	$R^I(u)$	$R^O(u)$	$R^I(u) \cap R^O(u)$
1	$\{1,4,7\}$	$\{1,4\}$	$\{1,4\}$
2	$\{2,5\}$	$\{2,5,8\}$	$\{2,5\}$
3	$\{3,6\}$	$\{3\}$	$\{3\}$
4	$\{1,4,7\}$	$\{1,4\}$	$\{1,4\}$
5	$\{2,5\}$	$\{2,5,8\}$	$\{2,5\}$
6	$\{6\}$	$\{3,6\}$	$\{6\}$
7	$\{7\}$	$\{1,4,7\}$	$\{7\}$
8	$\{2,5,8\}$	$\{8\}$	$\{8\}$

Notice that in the last column both 1 and 4 have the
entry $\{1,4\}$, and both 2 and 5 have $\{2,5\}$, while each of the
other points has the set of which it is the only member.

Looking at reachability as a relation on the set of
points, we see that it is reflexive, because of the trivial
walks. It is also transitive, because of Theorem 1.2. But
it is not in general symmetric, as shown by the fact that
the inreach and outreach of a point are not always the same.
Theorems 2.1 to 2.3 are little more than restatements of
Theorem 1.2 in the light of the new definitions 2.1 to 2.4,
and we omit the proofs.

THEOREM 2.1   (a) $R^O(v) \subseteq R^O(u)$ if and only if v is
reachable from u.
                (b) $R^I(u) \subseteq R^I(v)$ if and only if v is
reachable from u.

DEFINITION 2.3 If u and v are points in a digraph D,
and each is reachable from the other in D, then we
say u is _equivalent_ to v.

THEOREM 2.2   Equivalence is an equivalence relation.
        (u is equivalent to u;   if u is equivalent to v, then
        v is equivalent to u;   if u is equivalent to v and v
        is equivalent to w, then u is equivalent to w).
Since equivalence is an equivalence relation, it partitions
the set of points into equivalence classes.

    DEFINITION 2.4   If u is a point in a digraph D then
    the set of points equivalent to u is called the
    component of u, and is symbolised by C(u).
    Since the components are the equivalence classes, the
    components defined by two points are either the same
    set or they have no point in common.

THEOREM 2.3   If u is a point in a digraph,

$$C(u) = R^O(u) \cap R^I(u)$$

    Applying this result to H of Figure 1.3 we see that
the last column of the table we calculated gives the
components:

        $\{1,4\}$, $\{2,5\}$, $\{3\}$, $\{6\}$, $\{7\}$, $\{8\}$.

    Although this method of construction of the component
is quite efficient if we desire to know only the components
to which some of the points belong, if we wish to know the
components for all the points we can roughly halve the
labour by exploiting either of the following theorems.

THEOREM 2.4   If u and v are two points then u is equivalent
to v if and only if $R^O(u) = R^O(v)$.

PROOF   Suppose $R^O(u) = R^O(v)$. As each point belongs to its
own outreach, the equality of the sets implies that each
belongs to the other's. Hence each is reachable from the
other and they are equivalent.

    Conversely, if u is equivalent to v, u is reachable
from v, so that $R^O(u) \subseteq R^O(v)$ by Theorem 2.1. In the same
way $R^O(v) \subseteq R^O(u)$. Hence $R^O(u) = R^O(v)$.   □

THEOREM 2.5   If u and v are two points then u is equivalent
to v if and only if $R^I(u) = R^I(v)$.

The PROOF follows exactly the lines of the previous proof.
    These theorems allow us to find the components by
finding only the outreach sets or only the inreach sets and
then searching to collect together those points which have
equal sets. These collections are the components.

    For example, consulting the inreach and outreach
columns for digraph H, we see that points 1 and 4 have the
same inreach and outreach, and so have 2 and 5, while every

other point has inreach and outreach unique to itself.
Therefore the components are $\{1,4\}$, $\{2,5\}$, $\{3\}$, $\{6\}$, $\{7\}$,
$\{8\}$.

2.3 Component Analysis

With a little practice it is easy enough to see the
inreach, outreach and components in a small digraph
presented as in the previous section, so that walks are easy
to check; but where the digraph is large, or drawn as in
Figure 2.1, we need less laborious methods. We also need
methods that work on digraphs presented as a successor table
table.
In this section we develop a mechanical method or
algorithm for calculating the components of a digraph. We
will do this in terms of the outreach sets: it can equally
be done in terms of the inreach sets. The algorithm
depends on the following two theorems.

THEOREM 2.6 If u,v,w are points in a digraph D such that
$v \in R^O(u)$, and w is a successor of v, then $w \in R^O(u)$.

PROOF Since $v \in R^O(u)$ there is a walk from u to v. As
w is a successor of v, $\langle v,w \rangle$ is a walk from v to w. Thus
there is a walk from u to w, and $w \in R^O(u)$.   □

THEOREM 2.7 If u,w are points of a digraph D such that
$w \in R^O_O(u)$, then either $w = u$ or there exists a point
$v \in R^O(u)$ such that w is a successor of v.

PROOF Since $w \in R^O(u)$, there is a walk with u as first
and w as last point. Either $u = w$ or the walk has length at
least 1. Then the walk contains a last arc (v,w) for some
v. There is thus a walk with u as first point and v as last
point, and w is a successor of v.   □
We calculate the outreach of a point by building up
the set. Starting at the point itself we continually add
the successors of the points already in the set until we
reach a situation where all the successors of every point in
the set are also in the set. The process is then complete.
As the point set of the digraph is finite, the process
must terminate.

TO CALCULATE THE OUTREACH OF A POINT u

1. Draw up the digraph in successor table form.
2. Commence the list for $R^O(u)$.
3. Write u in the list.
4. Put a mark under u in the list.

5. Add to the list any successor of the last marked
    point which is not already in the list.
6. If the last point in the list has been marked go
    to 8. Otherwise go to 7.
7. Mark the first unmarked point in the list and go
    to 5.
8. End.

As an example we construct the outreach of the point
a in Figure 2.1, showing the various steps.

Step
1   Point   Successors

         a   b
         b   p,e
         c   d,k,m
         d   k
         e   m
         f   c
         g   e,p
         h   a,g
         j   d
         k   j
         m   e
         n   b
         p   h

2   List now reads
3   a
4   a̲
5   b is the only successor of a   ab̲
6   Last point not marked
7   Mark   first unmarked point   ab̲
5   Add successors of b   abp̲e
6   Last point not marked
7   Mark p   abp̲e
5   Add h, successor of p   abp̲eh
6   Last point not marked
7   Mark e   abpeh
5   Add m, successor of e   abp̲ehm
6   Last point not marked
7   Mark h   abp̲ehm
5   As a is already in list, add
    g only   abp̲ehmg
6   Last point not marked
7   Mark m   abp̲ehmg
5   e is already in list
6   Last point not marked

7 Mark g   abpehmg
5 e,p already in list
6 Last point marked
8 End.

    In order to find the components we carry out this
calculation for each point in turn. The marking of the
points in the description above is only a working device
and an aid to description. Physical marking can be omitted
as soon as the technique has been mastered.

TO CALCULATE THE COMPONENTS OF A DIGRAPH

1. Draw up a table with columns headed Point,
    Successors, Outreach, Component.
2. Enter the points in the 'Point' column and their
    successors in the 'Successors' column.
3. In the 'Outreach' column enter the outreach of
    each point, calculated as above.
4. Searching the 'Outreach' column identify the
    distinct outreach sets and collect their points to
    form the components.
5. End.

For example the completed table for the digraph of Figure
2.1 is:

Point	Successors	Outreach	Components
a	b	a b p e h m g	a b g h p
b	p e	b p e h m a g	
c	d k m	c d k m j e	c
d	k	d k j	d j k
e	m	e m	e m
f	c	f c d k m j e	f
g	e p	g e p m h a b	
h	a g	h a g b e p m	
j	d	j d k	
k	j	k j d	
m	e	m e	
n	b	n b p e h m a g	n
p	h	p h a g b e m	

The process of identifying equal outreaches is conveniently
broken into several stages. Most obviously, equivalent
points must have the same number of points in their
outreaches. This enables a first separation to be made.
Next the outreach of each point need only be compared with
the outreaches of points in its own outreach. Further,
this comparison can be reduced to the question: Point v is
in the outreach of u; is u in the outreach of v? If the

answer is 'Yes' they are equivalent; if it is 'No', they
are not. If an eye is kept open for this during the
construction of the outreaches, the construction of those
sets can sometimes be speeded up.

The components of the digraph of Figure 2.1 are thus
$C_1 = \{a,b,g,h,p\}$, $C_2 = \{d,j,k\}$, $C_3 = \{e,m\}$, $C_4 = \{c\}$
$C_5 = \{f\}$, $C_6 = \{n\}$.

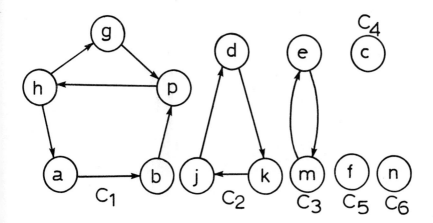

FIGURE 2.3

Each component is a subset of the point set and
therefore defines a subdigraph, called a <u>component
subdigraph</u> in formal statements, but we will call both set
and subdigraph just component in informal work. The
component subdigraphs for the present example are shown in
Figure 2.3.

In terms of our biological analogy we have now
displayed the organs of the digraph. In Section 2.5 we
will investigate the interrelation between the organs.

2.4 Exercises

S 1. Find the inreach and outreach sets and components for
digraph A of Figure 1.3.

2. Find the inreach and outreach sets and components for
digraph D of Figure 1.3.

S 3. Let $V = \{a,b,c,d,e,f,g,h\}$
$A = \{(a,b),(c,a),(c,d),(c,e),(f,d),(f,h),(g,h),$
$(h,e)\}$.

This digraph is interpreted by taking the points to
represent people, and an arc (u,v) to mean that 'u
can persuade v'. A measure of a person's power is the
number of people he can persuade, either directly or
through one or more intermediaries. How is the set
of such people described in digraph terms? Which of
these people is the most powerful?

S 4. Find the inreach for each course in the example at the
beginning of section 2.2, and shown in Figure 2.2. A
valid combination of courses must contain the inreach
of every course that belongs to it, and must therefore
be a union of inreaches. Find three valid combinations
which consist of exactly five courses.

S 5. What is the relation between the inreach of a point in
a digraph and its inreach in some partial digraph?
What is the relation between the components in the two
cases?

6. What is the relationship between the inreach of a
point in a digraph and its inreach in some subdigraph?
What is the relationship between the corresponding
components?

S 7. Carry out a component analysis for digraph E of
Figure 1.3.

8. Carry out a component analysis for digraph G of
Figure 1.3.

S 9. (Sociogram). A group of boys were asked to choose
their two best friends in the group. The choices were:

Arthur:	Barry, Charles
Barry:	Henry, Arthur
Charles:	Barry, Daniel
Daniel:	Arthur, Michael
Edward:	Frank, Gary
Frank:	Edward, Ivan
Gary:	Edward, Frank
Henry:	Ivan, Kenneth
Ivan:	Henry, Kenneth
John:	Arthur, Michael
Kenneth:	Henry, Ivan
Michael:	Edward, Gary

Describe the situation by means of a digraph and find
its components. Suggest an interpretation of the
results.

10. Three judges were asked to rank ten brands of cheese,
each constructing his own list. The lists were:

Judge	1st	2nd	3rd	4th	5th	6th	7th	8th	9th	10th
A	k	m	n	p	q	r	s	t	u	v
B	m	k	p	q	n	t	s	r	v	u
C	m	k	n	q	p	t	s	v	r	u

Construct a digraph which incorporates all these
rankings. Carry out a component analysis and interpret
the result.

2.5 Condensation

In order to represent the interrelationships between
the components we define a new digraph, the condensation
digraph, whose points correspond to the components of the
original digraph, and whose arcs indicate the existence of
arcs which begin in one component and end in another.

DEFINITION 2.5 Let $D = (V,A)$ be a digraph with
components C_1, C_2, ..., C_k.
Let $V^* = \{C_1, C_2, ..., C_k\}$
$\quad A^* = \{(C_i,C_j): \quad i \neq j;$ there exist $v \in C_i$ and
$\qquad\qquad\qquad\qquad w \in C_j$ such that
$\qquad\qquad\qquad\qquad (v,w) \in A\}$
Then $D^* = (V^*,A^*)$ is called the condensation of D.

We refer once more to the example of Figure 2.1. We
found its components in the previous section. Many of its
arcs join points in the same component and are irrelevant
here. The remainder are (b,e) and (g,e) which give rise to
(C_1,C_2); (c,d) and (c,k) which give rise to (C_4, C_2);
(c,m), giving (C_4,C_3); (f,c), giving (C_5,C_4) and (n,b),
giving (C_6,C_1). The condensation digraph is therefore:
$\quad V^* = \{C_1,C_2,C_3,C_4,C_5,C_6\}$
$\quad A^* = \{(C_1,C_3),(C_4,C_2),(C_4,C_3),(C_5,C_4),(C_6,C_1)\}$,
and can be drawn as in Figure 2.4.
The condensation digraph indicates the relationships
between components, and so for many purposes is the
appropriate digraph to consider. When we are using the
components and condensation to help us interpret the
original digraph, however, we may be interested in the
individual arcs between components and their beginning and
end points.

DEFINITION 2.6 If (u,v) is an arc in a digraph D,
with u and v belonging to different components
C_i, C_j respectively of D, we say that u is an outgate

c

of C_i and v is an <u>ingate</u> of C_j. A point which is both
an ingate and an outgate will be called a <u>bigate</u>.

    Thus in our example b and g are outgates of C_1: b is
also an ingate, and is therefore a bigate. In C_2 both d and
and k are ingates and there is no outgate. In C_3 both e
and m are ingates and there is no outgate. The point c is a
a bigate for the component to which it belongs, and f and n
are outgates for their respective components.

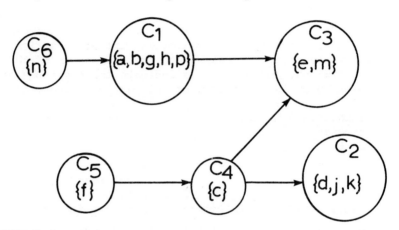

FIGURE 2.4

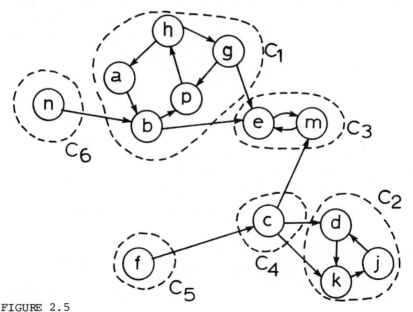

FIGURE 2.5

The whole digraph can now be redrawn, arranging each
component appropriately and laying them out in the relative
positions of Figure 2.4.   Figure 2.5 is the redrawn digraph.
     There are two situations when this analysis gives no
simplification of the digraph structure.   It may be that all
points are equivalent.   There is then only one component,
so that the component subdigraph is the original digraph,
and the condensation consists of a single point, and no
arcs.   Such a digraph is called <u>strong</u>, and sections
2.9 and 2.10 deal with strong digraphs.
     At the other extreme each component may consist of a
single point.   Then the condensation digraph is isomorphic
to the original digraph, with {u} replacing u in each case.
Such digraphs are termed <u>acyclic</u>.   In many applications the
digraphs must automatically be acyclic, so the whole of
Chapter 3 is devoted to these digraphs and there is much
reference to special acyclic digraphs (such as rooted trees)
in later chapters.
     Since it happens that the components of a digraph are
always strong, and the condensation always acyclic, we can
never use component analysis more than once.
     The condensation is in fact simply the most important
example of a wider operation of constructing one digraph
from another.   Instead of the earlier stages of the analysis
producing the subdigraphs, they can be defined in some
other way, either by a different digraph-theoretical
equivalence relation or by virtue of the meanings attached
to the points.
     In Section 2.2 we used an example where the points
represented courses.   The letters represented the subject,
and the number the level.   If we are interested in the
relation between subjects only, we may partition the
points:
     C = {C1}   ;   LA = {LA1,LA2,LA3}   ;   NS = {NS1}   ;
        AA = {AA2,AA3}   ;   MP = {MP2,MP3}.
     These sets become the points of a new digraph, whose
arcs indicate that some course in one set is a prerequisite
for some course in another, as in Figure 2.6.

DEFINITION 2.7 Let $D = (V,A)$ be a digraph, and let
$\overline{P} = \{U_1,U_2,\ldots,U_k\}$ be a partition of V.   Let
$B = \{ (U_i,U_j): \ i \neq j$ and there exists $(v,w) \in A$ such that
                $v \in U_i$ and $w \in U_j\}$
     Then (P,B) is called a <u>factor digraph</u> of (V,A).
The condensation of a digraph is thus its factor digraph
relative to its component partition.

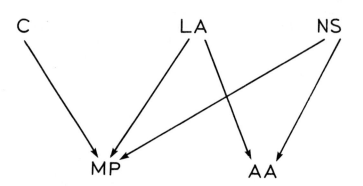

FIGURE 2.6

2.6 Components and closed walks

The two previous sections laid out methods by which
the components of a digraph may be computed mechanically.
Very often it is possible to find the components by
inspection. The purpose of the present section is to
develop inspection methods and at the same time
investigate further the nature of components.
First, just as reachability is concerned with the
existence of walks, so equivalence can be shown to be
concerned with the existence of closed walks.

THEOREM 2.8   If u and v are two points in a digraph there
is a closed walk with u as its first and last point and
passing through v, if and only if u and v are equivalent.

PROOF   If there is a closed walk with u as its first and
last point and passing through v, then this walk may be
split into two walks, P with first point u and last point
v, and Q with first point v and last point u.   Thus u and
v are equivalent.
Conversely if they are equivalent walks P and Q as
described above exist, and their join PQ is a closed walk
with u as first and last point and passing through v.   □
Hence any two points on the same closed walk are
equivalent. Collecting such closed walks we can build up
the components by combining sets of points on closed walks
which have points in common.

This leaves the problem of demonstrating non-equivalence. This can be done by dividing the points of a digraph into two sets, such that either there are no arcs between points in different sets, or they all run in the same direction.

DEFINITION 2.8   Let D = (V,A) be a digraph, and let V = W $\cup$ X, with W $\cap$ X = $\emptyset$. If there are no arcs which begin in W and end in X, we say (W,X) is a <u>section</u> of D.
This name is chosen because of the parallel with the Dedekind section of the rational numbers. It is an ordered pair because of the asymmetry in the roles of W and X.

THEOREM 2.9   If (W,X) is a section of D, no component of D has points in both W and X.

PROOF   We show in fact that no point in X is reachable from any point in W.
For suppose $u_0 \in$ W and $u_m \in$ X such that there is a walk

$$\langle u_0, u_1, \ldots, u_m \rangle$$

Then each u belongs to either W or X, so there is a first u which belongs to X. We know further that i $\neq$ 0. Then there is a point u_{i-1} in the walk and $u_{i-1} \in$ W. But (u_{i-1}, u_i) is an arc of the walk and therefore of D, contrary to hypothesis. Hence there is no such walk.
From this it follows that no point of X is reachable from any point of W, so no point of X is equivalent to any point of W, as was required.   □
By combining several sections we can partition the point set into smaller and smaller parts, each containing whole components. At the same time components are being built up 'from beneath'. We may formalise a set of rules for this process as follows:
(a) If $\langle u,v,\ldots,z,u \rangle$ is a closed walk, then {u,v,...,z} is a subcomponent.
(b) The union of two subcomponents with a common member is a subcomponent.
(c) If (W,X) is a section then W and X are supercomponents.
(d) The intersection of two supercomponents (unless empty) is a supercomponent.
(e) A set which is both a subcomponent and a supercomponent is a component.
Although the method is designed for informal use and not normally written down we exemplify the method by reference to Figure 2.7.

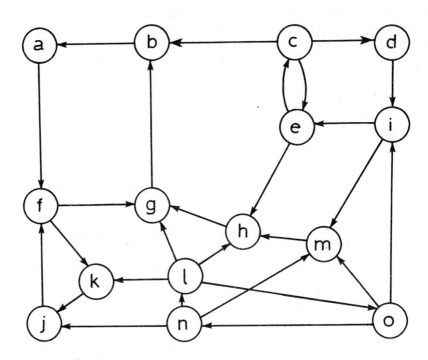

FIGURE 2.7

We notice the cycles $\langle c,e,c \rangle$, $\langle f,k,j,f \rangle$, $\langle a,f,g,b,a \rangle$.
So $\{c,e\}$, $\{f,k,j\}$ and $\{a,f,g,b\}$ are subcomponents. As the
last two have a common member their union is also a subcomponent. Mentally dividing off this set $\{a,f,g,b,k,j\}$,
we notice that although arcs enter the set, no arcs leave.
Hence this is the left-hand member of a section, of which
the other member is $\{c,d,e,i,h,m,l,o,n\}$. As $\{a,f,g,b,k,j\}$
is both a subcomponent and a supercomponent it is a
component, and may now be ignored.

Within the other supercomponents we find a cycle
$\{l,o,n,l\}$, so $\{l,o,n\}$ is a subcomponent. We notice that it
is also the right hand member of a section, so it is a
component.

We already have $\{c,e\}$ as a subcomponent, and we may
observe that the intersection of the right hand member of
the first section with the left hand member of the second
is $\{c,d,e,i,h,m\}$, which is therefore a supercomponent.
Another section yields $\{c,d,e,i,l,o,n\}$ as a supercomponent,
separating off $\{c,d,e,i\}$. We then observe the closed walk
$\langle c,d,i,e,c \rangle$, so that $\{c,d,e,i\}$ is a component. The method
can be continued to yield the rest of the components.

If pursued with sufficient vigour the method will
always produce the components, but the more complex the
digraph the harder it will be to find the closed walks, and
even more so the sections. It is not recommended for
digraphs much more complicated than that shown here.   For
complicated digraphs the algorithm should be used.
Other short-cuts are indicated by some of the
exercises in Section 2.8.

2.7   Weak Components

Looking at digraph H of Figure 1.3, the most obvious
thing about it is that it falls into three 'pieces', with
no arc running in either direction between points in
different pieces.   The component analysis does not yield
these three pieces but six components, which are each
contained in one piece.
These 'pieces' are called weak components.   In a well-
drawn digraph they are immediately obvious, but they are
much less important than the components we have described.
In many texts the sets we have called just 'components' are
called strong components.   A digraph with only one weak
component is called weakly connected or weak, while as we
have said, a digraph with one strong component is called
strong, or strongly connected.
In most books (for example [Harary]) the definition
of weak components involves a generalisation of the concept
of a walk.   We, on the other hand will adapt the sub-
component/supercomponent approach of the previous section.
We use the rules:
- (a) For each point u, $\{u\}$ is an inner set.
- (b) If (u,v) is an arc, $\{u,v\}$ is an inner set.
- (c) The union of two inner sets with a point in
  common is an inner set.
- (d) The set V of all points in the digraph is an outer
  set.
- (e) If $V = W \cup X$, with $W \cap X = \emptyset$, such that no arc
  begins in either set and ends in the other, then
  W and X are outer sets.
- (f) The intersection of two outer sets is an outer
  set.
- (g) The weak components are precisely the sets which
  are both inner and outer.

The weak components form a partition of V: results about
weak components are set in Exercises 2.8.

2.8 Exercises

S 1. Find the components and the condensation of the
digraph A of Figure 1.3.

S 2. Find the components and condensations of the digraphs
D and E (=D^C) of Figure 1.12. Make and prove
conjectures about the components and condensations of
converse digraphs.

S 3. B.A. students studying certain major subjects also
study various minor subjects, as shown by the table
below, based on a sample of University of Canterbury
students. Consider the digraph whose points are the
various subjects listed, and whose arcs join major to
minor subjects. We have arbitrarily restricted each
subject to its three most popular minors. Note that
there are no students in the sample with (13) AMST
(American Studies) or (14) MATH as their major
subject.

Find the components and the condensation digraph
and the various gates. As often happens there is one
large component and the rest consist of only one or
two points.

	Major	Minors
1	ECON	MATH,POLS,SOCI
2	EDUC	ENGL,PSYC,SOCI
3	ENGL	AMST,HIST,FREN
4	GEOG	EDUC,HIST,SOCI
5	FREN	ENGL,GRMN,HIST
6	GRMN	ENGL,HIST,FREN
7	HIST	ENGL,GEOG,SOCI
8	JAPA	ECON,ENGL,POLS
9	MUSI	EDUC,GEOG,SOCI
10	POLS	ECON,HIST,SOCI
11	PSYC	EDUC,ENGL,SOCI
12	SOCI	EDUC,ENGL,PSYC

4. Suppose we partition the subjects mentioned in
Question 3: {ECON,MATH}, {EDUC,PSYC,SOCI}, {ENGL,AMST},
{GEOG,HIST}, {FREN,GRMN}, {JAPA}, {POLS}, {MUSI}.
Construct the corresponding subdigraphs and the factor
digraph.

S 5. Prove that if u and v are points of a digraph D with
v reachable from u, and if C(u) and C(v) are their
respective (different) components, then C(v) is
reachable from C(u) in the condensation D*.

* 6. Prove the converse of the above result, namely that if
C(v) is reachable from C(u) in D*, then v is reachable
from u in D. Use this result to prove that the

condensation of any digraph is acyclic, in terms of
the definition used in this section.

S 7. Find the factor digraph of Figure 1.9 in which the
points are partitioned:
CO_2 = {CO_2 in air, CO_2 in water},
geological forms = {fossil fuels, limestone}
living forms = {green plants, animals, bacteria, fungi}
dead forms = {dead matter}.

S 8. Complete the finding of the components in the example
at the end of section 2.6 shown in Figure 2.7.

9. Suppose u and v are two points known to be in the same
component, and that a walk has been found from u to v.
Show that all points on this walk are also in the same
component. Develop a new rule about subcomponents to
incorporate this result.

10. Show that every set consisting of a single point is a
subcomponent, and the set of all points is a supercomponent according to our rules.

*11. Prove that the weak components partition the point set.

S 12. Prove that if W(u) is the weak component to which u
belongs, then $R^I(u)$, $R^O(u)$ and C(u)
are all subsets of W(u).

13. Show that in a symmetric digraph
$$W(u) = R^I(u) = R^O(u) = C(u).$$

* 14. Let D be a digraph and C a component of D. Let the
subdigraph defined by (V \ C) have components
$C_1, C_2, \ldots, C_k$. Prove that the components of D are
$C, C_1, C_2, \ldots, C_k$.

15. Prove that if id(u) = 0 or od(u) = 0, then C(u) = {u}.

S 16. Suppose that {u,v} is a component of some digraph.
Prove that (u,v) and (v,u) are both arcs of the
digraph.

17. Figure 2.8 shows a digraph. Use the methods of
section 2.6 to find the components. Find also the
gates of each component.

S 18. Figure 2.8 has one source and three sinks. Find these
points. For each sink find the shortest path from
the source to that sink. Is it true that in every
digraph with one source and several sinks there is a
walk from the source to each sink?

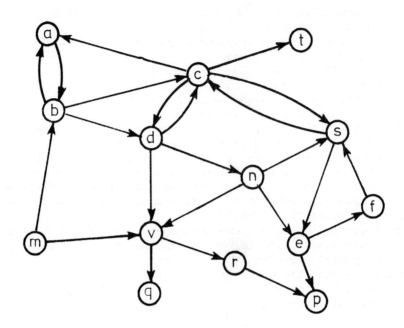

FIGURE 2.8

S 19.Morgan et al (Behaviour, 1976) quoted data on the times
each pair out of eleven chimpanzees spent together.
The list below gives the three companions with which
each ape spent most time. Interpreting this as the
successor table of a digraph, find the components and
the condensation.

Ape	Companions		
Ch	Mk	Hh	Hg
Ev	Hg	Mk	Hm
Fb	Fg	Hg	Mk
Fg	Fb	Mk	Hg
Go	Mk	Hg	Hm
Hh	Ch	Mk	Hg
Hg	Hm	Mk	Ly
Hm	Ly	Mk	Hg
Ly	Hm	Mk	Hg
Mk	Hg	Hm	Ly
St	Hg	Ly	Mk

2.9 Strong Digraphs

We have already mentioned strong digraphs informally.
In this section we define them formally and derive a few
results.

DEFINITION 2.9 A digraph is strong if every point can
be reached from every point.

THEOREM 2.10 If C is a component subdigraph of a digraph
D, then C is strong.

PROOF If u and v are two points in C, then each can be
reached from the other by walks in D: what we have to show
is that they can be reached from each other by walks in C.
Let P be a walk in D from u to v. Let Q be a walk in D
from v to u. Then PQ is a closed walk in D, and as we saw
in Section 2.6, all the points on a closed walk are
equivalent. Hence every point on PQ is a point of C. Thus
v is reachable from u in C, and C is strong.   □
     We have incidentally encountered another criterion for
a digraph to be strong.

THEOREM 2.11 A digraph D is strong if and only if given
any two points u and v there is a closed walk beginning and
ending at u and passing through v.
     This theorem is a special case of Theorem 2.8.   □

THEOREM 2.12 A digraph is strong if and only if it has a
spanning closed walk.

PROOF Suppose (V,A) is a strong digraph. Let
$V = \{u_1, u_2, \ldots, u_p\}$. Then as there is a walk from each
point to each other point there are walks $W_1, W_2, \ldots, W_{p-1}$
such that the first point of W_i is u_i
and its last point is u_{i+1}, and a walk W_p whose first
point is u_p and last point u_1. Then

$W_1 W_2 W_3 \ldots W_p$ is a spanning closed walk.
     (A closed walk so generated will often be
unncessarily long. It is not, however, generally possible
to find a spanning cycle).
     The converse of the proof follows by observing that
if v follows u in the spanning closed walk then we can
extract a walk from u to v. On the other hand if v
precedes u, there is a walk from u to the last point of the
spanning walk, and a walk from that point (as first point)
to v. These can be joined to make a walk from u to v.   □

A digraph may of course be shown to be strong by ordinary component analysis. Alternatively it may be shown by finding a spanning closed walk. The subcomponent method can also be used, especially if the digraph is presented as an arrow diagram.

Strong digraphs occur naturally in many applications. The carbon cycle shown in Figure 1.9 is strong, and so are the digraphs of road networks, Figures 1.8 and 1.15 and the energy conversion digraph, problem 1.10.10. Others are indicated in Exercises 2.11.

2.10 Distance calculations

We introduced the concept of distance in Section 1.7, and proved the triangle inequality in Section 1.8. In this section we provide an algorithm for calculating distances and also apply distance criteria to the analysis of strong digraphs. Calculation of distances will receive further attention in Chapter 8.

In order to calculate distances it is necessary only to make a small modification to the component analysis table. At the cost of little extra work we can simultaneously calculate outreach and distance, for the members of the outreach of u are precisely the points to which distances from u can be measured.

We modify the outreach column by dividing it into columns corresponding to the numbers $0,1,2,\ldots,p-1$, the possible distances from each point, though generally far fewer columns are actually needed. In calculating the outreach of u, u goes in column 0 and if the marked point at any stage is in column i, its successors not already listed go into column i+1. For the example the table for Figure 2.1 is:

Point	Successors	0	1	2	3	4	5	 12
a	b	a	b	p,e	h,m	g		
b	p,e	b	p,e	h,m	a,g			
c	d,k,m	c	d,k,m	j,e				
d	k	d	k	j				
e	m	e	m					
f	c	f	c	d,k,m	j,e			
g	e,p	g	e,p	m,h	a	b		
h	a,g	h	a,g	b,e,p	m			
j	d	j	d	k				
k	j	k	j	d				
m	e	m	e					
n	b	n	b	p,e	h,m	a,g		
p	h	p	h	a,g	b,e	m		

It will be observed that as the column 0 contains only the point itself, and the 1 column contains just the successors of the point, two columns may be omitted.

From the above table the distances may be read off, and the distance table, which is also a reachability table constructed.

To

From	a	b	c	d	e	f	g	h	j	k	m	n	p
a	0	1		2		4	3				3		2
b	3	0		1		3	2				2		1
c			0	1	2				2	1	1		
d				0					2	1			
e					0						1		
f			1	2	3	0			3	2	2		
g	3	4		1			0	2			2		2
h	1	2		2			1	0			3		2
j				1					0	2			
k				2					1	0			
m				1							0		
n	4	1		2		4	3				3	0	2
p	2	3		3		2	1				4		0

With non-strong digraphs reachability considerations dominate, but in a strong digraph there are no problems with existence of distances, and we can make more use of the actual values.

DEFINITION 2.10   The diameter of a digraph is the maximal value of $d(u,v)$ over all pairs of points. A diametral path is a path of shortest length linking points whose distance apart is the diameter of the digraph.

The second part of the definition is possible because we have already shown in Theorem 1.7 that for any u and v for which $d(u,v)$ is defined there is a path of length $d(u,v)$ from u to v.

In the example used in section 1.7 (page 26), whose table is repeated on page 66, the diameter is

$$4 = d(e,c) = d(e,d).$$

Figure 1.6 easily yields the diametral paths $\langle e,f,a,b,c\rangle$ and $\langle e,f,a,b,d\rangle$. In this case each pair of points defines a unique diametral path, but this need not be the case.

In the construction of diametral paths the methods of rooted trees (Chapter 5) and networks (Chapter 6) are valuable.   The process is facilitated if the

distance table is known in some form, for the points
along a diametral path are successively at distances 0,1,2,
3,... from the first point, and each is a successor of the
point which precedes it in the path. This usually reduces
the exploration to modest proportions.

When the points represent people, and the arcs their
interaction, or the points are geographical and arcs
represent physical access, or in several other situations
the concept of central persons or places may have
significance. There are several methods for defining
centres, but the easiest ones to compute employ the row and
column sums of the distance table.

DEFINITION 2.11   The outsum of point u in a strong
digraph is the sum of the distances to all the points
of the digraph:
$$s^0(u) = \sum_{v \varepsilon V} d(u,v).$$

This is calculated as the sum of the numbers in the
row of u.   The insum of point u in a strong digraph is
the sum of distances from all the points of the
digraph to u.
$$s^I(u) = \sum_{v \varepsilon V} d(v,u).$$

This is calculated as the sum of the numbers in the
column of u.

The totsum of point u in a strong digraph is the sum of its
insum and outsum.

An outcentre of a strong digraph is a point of minimum
outsum.

An incentre of a strong digraph is a point of minimum
insum.

A totcentre of a strong digraph is a point of minimum
totsum.

Thus we add to the distance table one new column and
two new rows for these sums. The table for the example
given in Section 1.7 and referred to above is now:

	a	b	c	d	e	f	s^0
a	0	1	2	2	1	2	8
b	1	0	1	1	2	3	8
c	2	3	0	1	2	3	11
d	1	2	3	0	1	2	9
e	2	3	4	4	0	1	14
f	1	2	3	3	2	0	11
s^I	7	11	13	11	9	11	61
tot	15	19	24	20	22	22	

The sum of the insums and the sum of the outsums should be
calculated as a check. As each is the sum of all the table
entries the totals should agree.

In this case there are two outcentres, a and b, and
one incentre, a again. When any point is both an outcentre
and an incentre it must be a totcentre, but it is possible
to have totcentres that are neither outcentres nor
incentres. (See exercise 2.11.20).

It is often desired to have also a measure of central
tendency. In a strong digraph with p points the minimum
possible insum and outsum are p - 1: a measure of
centrality for any point can be found by dividing p - 1
by the insum or outsum, or 2p - 2 by the totsum, as
appropriate to the application.

The minimum value of the sum of all the table entries
is p(p - 1), when all non-diagonal entries are 1. This
occurs when the digraph is complete. If we divide p(p - 1)
by the sum of the distance table entries we obtain a
measure of the digraph as a whole, which is 1 when the
digraph is complete and less otherwise. Some authors have
interpreted this value, or some simple function of it as a
measure of centralisation. In this view a complete
digraph is completely centralised. We however prefer to
see it as a measure of compactness, and call it the
compactness ratio but in all applications it matters less
what the number is called than the use made of it.

2.11 Exercises

S 1. Prove that the road network, Figure 1.8 is strong.
  2. Prove that the road network, Figure 1.15 is strong.
  3. Prove that the carbon cycle digraph, Figure 1.9 is
     strong.
S 4. Which of the digraphs in Figure 1.3 are strong?
S 5. Prove that a complete digraph is strong.
S 6. Show that the converse of a strong digraph is strong.
S 7. Prove or disprove the conjecture that every
     subdigraph of a strong digraph is strong.
  8. Prove or disprove the conjecture that every partial
     digraph of a strong digraph is strong.
S 9. What is the minimal number of arcs possible in a
     strong digraph with p points?
S*10.What is the maximal number of arcs in a digraph which
     is not strong? Experiment with digraphs having
     2,3,4,5 points first.
 * 11.Suppose we call two points 'concyclic' if there is a
     cycle through both of them. Prove that the concyclic

relation is not an equivalence relation on a strong
digraph.

12.Show that in any strong digraph there is a closed walk
which includes every arc at least once.

13.Construct the distance table for the Carbon Cycle,
Figure 1.9, and find its centres and its compactness
ratio.

* 14.What is the maximum possible value of the outsum of
any point in a strong digraph with p points?

15.What is the maximum possible diameter of a strong
digraph with p points?

S 16.Find the insum, outsum and totsum of the parking
building (PB) in Figure 1.8.

S 17.Suppose that a strong digraph remains strong upon the
removal of a certain arc. Show that the compactness
ratio is reduced.

* 18.Construct several strong digraphs, calculate their
compactness ratios and arrange them in order of
increasing compactness. Does this order agree with
your conception of compactness? Can you find a
better indicator?

* 19.The commonest ways in which the chromosome numbers of
a plant can differ from those of its parents are by
doubling, or by reduction of 1. Construct the
digraph of all integers from 1 to 20 (an arbitrary
upper bound) as points, and all arcs of the forms
$(n, n-1)$ and $(n, 2n)$ between these points. Prove that
the digraph is strong. Does it have a spanning cycle?
Construct the distance table, find the diameter and
diametral paths.

S 20.If $V = \{a, b, c, d, e, f\}$ and $A = \{(a, b), (a, c), (a, d), (a, e),$
$(a, f), (b, c), (c, f), (d, a), (d, c), (d, e), (d, f), (e, d), (e, f),$
$(f, e)\}$ show that (V, A) has unique incentre, outcentre
and totcentre and that they are all different.

21.The table below lists nine people who work together,
and the persons to whom each will pass on information
of a particular kind that he hears. Consider this as
defining a digraph. Prove that the digraph is strong
and find the distance table. Find incentres and
outcentres. Interpret the meaning of incentre and
outcentre in this situation.

person	tells				
h	b	r	j		
b	h	j	d		
r	w	p	h		
s	h	b			
p	r	w	j	s	b
a	d	j			
d	j	h	s		
j	d	h	a		
w	p	r			

CHAPTER THREE

ACYCLIC DIGRAPHS

3.1 Definitions and tests

We introduced the term acyclic digraph in Section 2.5 to describe digraphs in which each component consists of a single point. Our formal definition is different, but is shown at once to refer to the same digraphs.

> DEFINITION 3.1 A digraph is acyclic if it has no non-trivial closed walks. A digraph is cyclic if it has at least one non-trivial closed walk. Thus every digraph is either cyclic or acyclic, and no digraph is both.

THEOREM 3.1 A digraph is acyclic if and only if all its components consist of single points.

PROOF This follows at once from Theorem 2.8. □
Component analysis will certainly prove that a digraph is acyclic, but as a test it is excessively slow. We will develop faster and easier methods, but first we need some theorems and definitions.

THEOREM 3.2 In an acyclic digraph every walk is a path.

PROOF Suppose W is a walk but not a path. Then at least one point of W must occur more than once. Let x be such a point. Then that section of the walk W between the first and second occurrences of x is a closed walk, and is non-trivial. Therefore the digraph is not acyclic. □

THEOREM 3.3 Every acyclic digraph has at least one sink.

PROOF Suppose that D is a digraph with p points, and that none of them have outdegree zero. We shall prove that this assumption implies that D must have at least one non-trivial closed walk.

Let u_1 be any point of D. As u_1 has non-zero outdegree it has at least one successor. Let u_2 be such a successor. In the same way u_2 must have a successor; call it u_3. In this way we build up a walk $\langle u_1, u_2, u_3, \ldots, u_p \rangle$. If the walk does not contain a non-trivial closed walk all these points must be different. So these are all points in the digraph. What then of the successor of u_p? If any of $u_1, \ldots, u_{p-1}$ is a successor of u_p we have a closed walk. Hence u_p has outdegree zero. We have reached a contradiction.   □

THEOREM 3.4 Every acyclic digraph has at least one source. The PROOF follows the same lines as that of the previous theorem.

These two theorems, together with the definition make useful tests: any digraph in which we can find a non-trivial closed walk, or in which we cannot find a point with indegree zero and a point of outdegree zero, is cyclic. We stress however that a digraph may have points of indegree zero and points of outdegree zero and yet not be acyclic.

THEOREM 3.5 Every subdigraph of an acyclic digraph is acyclic.

PROOF Suppose that D is a digraph and E is a subdigraph of D. If E is cyclic it has a non-trivial closed walk Q. But then Q is also a closed walk in D, hence D is cyclic. Hence an acyclic digraph which has a cyclic subdigraph must itself be cyclic, a contradiction which produces the desired result.   □

THEOREM 3.6 Every partial digraph of an acyclic digraph is acyclic.
The PROOF follows exactly the same lines as the previous proof.   □

THEOREM 3.7 If (V,A) is a digraph, and if there exists a function f from V to the real numbers such that if $(u,v) \in A$, then $f(u) < f(v)$, then (V,A) is acyclic.

PROOF Suppose (V,A) has a non-trivial closed walk
$$\langle u_1, u_2, u_3, \ldots, u_m, u_1 \rangle .$$

Then if f exists,
$$f(u_1) < f(u_2) < \ldots < f(u_m) < f(u_1)$$
But this implies that
$$f(u_1) < f(u_1)$$
which is false. Hence there can be no non-trivial closed walk in (V,A) and the digraph must be acyclic.   □

This theorem can be used as a criterion to establish that a digraph is acyclic. For example in Figure 2.2 the label of each point ends in a number and associates with each point a number with the properties required. In Figure 2.4 we can assign to each point the distance of the centre of its circle from the left edge of the page. If a digraph can be drawn so that its arcs all go from left to right, or down the page, or in some other consistent direction, then the digraph is acyclic.

We shall make use of two kinds of function which assign numbers to points in such a way that the number at the beginning of each arc is less than the number at its end: logical numberings in this section, and levels in a later section.

DEFINITION 3.2 A logical numbering on a digraph D with p points is a function n which assigns to each point u of D an integer n(u) so that:
(i) each of the integers 1,2,...,p is assigned to exactly one point;
(ii) if (u,v) is an arc, n(u) < n(v).

THEOREM 3.8 A digraph has a logical numbering if and only if it is acyclic.

PROOF Theorem 3.7 shows at once that if there is a logical numbering then the digraph is acyclic.

On the other hand, suppose that (V,A) is an acyclic digraph. By Theorem 3.4 it has at least one point with indegree zero. Choose such a point u and set n(u) = 1.

Now suppose that m < p points have been numbered. Let U_m be the set of points not numbered. Consider the subdigraph defined by U_m . This is acyclic by Theorem 3.5, so there is a point in this subdigraph with indegree zero. Assign to this point the number m+1. Continue until all points have been numbered.

As no point is numbered until all its precursors have been numbered, each arc has a smaller number at its beginning than at its end. This completes the proof.   □

A digraph will usually have many logical numberings, and for proof of acyclicity they are equally good.

Figure 3.1 shows all possible logical numberings of a
certain digraph with five points.

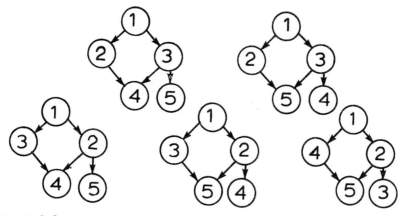

FIGURE 3.1

The construction used in the above proof can easily be
adapted to be a method for constructing a logical
numbering or proving alternatively that none exists. We
give two versions, one for arrow diagrams and one for
successor tables.

TEST FOR THE ACYCLIC PROPERTY: ARROW DIAGRAMS

1. If there is no point with indegree zero the
   digraph is not acyclic. End.
2. If there is a point with indegree zero, assign the
   number 1 to one such point.
3. If every unnumbered point has an unnumbered
   precursor, the digraph is not acyclic. End.
4. If there is an unnumbered point which has no
   precursors, or all of whose precursors have been
   numbered, assign the next integer to one such
   point.
5. If some point has not been numbered return to
   step 3.
6. If all points have been numbered, the digraph is
   acyclic, and a logical numbering has been obtained.
   End.

The flow chart for this test is Figure 3.2.
For the successor table version it is more convenient
to begin by assigning the number p and end by assigning 1.

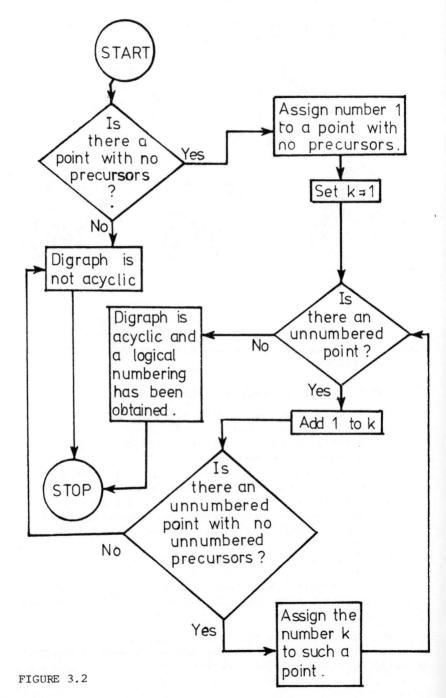

FIGURE 3.2

TEST FOR ACYCLIC PROPERTY: SUCCESSOR TABLE

1. If every point has a successor the digraph is not
   acyclic. End.
2. If there is a point which has no successor, assign
   the number p to such a point. Delete the name of
   this point wherever it occurs in the table as a
   successor.
3. If every unnumbered point has an undeleted
   successor the digraph is not acyclic. End.
4. If there is an unnumbered point whch has no
   undeleted successors, assign the next lower integer
   to one such point.
5. If some point has not been numbered return to
   step 3.
6. If all points have been numbered the digraph is
   acyclic and a logical numbering has been obtained.
   End.

The flow chart for this test is Figure 3.3.

3.2 Applications of Logical Numbering

Besides proving that a digraph is acyclic, itself a
necessary thing in many applications, the logical
numbering may be of value in its own right. In this
section we describe briefly some applications of the logical
numbering. A further application appears in Chapter 7 as
the first part of Activity Network Analysis.
(a) Cooking recipes and kitset instructions.
   The common feature here is that a set of instructions
   is written down to guide someone who has not
   performed the task before. The task must be broken
   down into a number of steps, so that the steps may be
   performed in a prescribed order. In designing these
   instructions it may be valuable to construct a
   digraph whose points are the various steps. Arcs will
   mean that one step has to be performed before another.
   If the digraph is not acyclic the steps must be
   rethought, or perhaps some of the arcs do not
   correspond to genuine restrictions. Each logical
   numbering corresponds to a possible order for the
   instructions. The actual choice of order may involve
   other considerations which have not been incorporated
   in the digraph.
(b) Course structure, lesson preparation and textbooks.
   Particularly in mathematics, but at times in most other
   subjects too, material must be presented in a course,

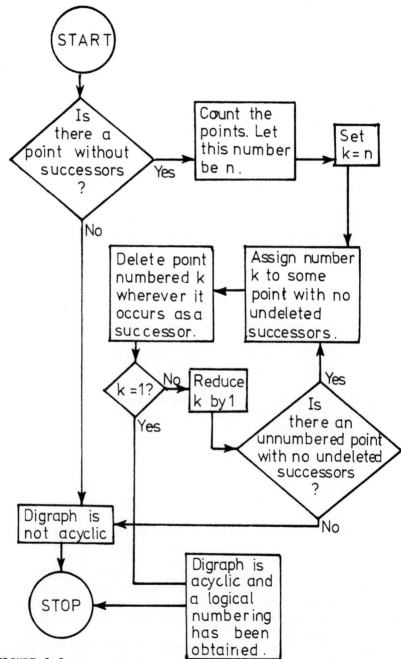

FIGURE 3.3

a single lecture, or a book or in some other way in an
order which is very largely determined by the logical
structure of the material. Yet it is not often
restricted to just one order; the subject branches
out, lines of argument divide and recombine. In such
circumstances it is useful to construct a digraph whose
whose points represent blocks of work, and whose arcs
represent precedence on grounds of logical structure
or some principle of teaching. Each logical numbering
is then a teaching order. Such digraphs were used in
the planning of parts of this book, such as the
previous section. We commend this method as an aid
but not as a master, recognising that as teaching is
not a purely logical activity, the edges of the
'blocks' are often not sharp, and will be modified
when the actual detail is written. By way of example,
Figure 3.4 shows the blocks for Section 3.1,
together with the direct quotations from earlier
sections. We have given the logical numbering of the
subdigraph whose points represent blocks of 3.1 itself
which corresponds to the actual order of material.

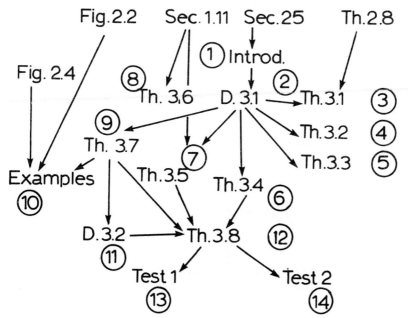

FIGURE 3.4

(c)   <u>Mathematical calculations</u>
In this application the logical numbering is less
important than in the other two, but the construction
of the digraph may be a useful aid.
Suppose we are given the following mathematical
problem:
     Eliminate r and s from the equations
        $n = a + r$; $a = r + s$; $r = s + 1$
     to obtain an expression of the form
        $x.a + y.n = 1$
We construct a digraph whose points represent equations.
Each new equation has as its precursors the equations
from which it was calculated. The problem in an
exercise of this type is not to get lost. The digraph
may help to avoid this. It may also be used in
teaching to display the argument in a form which is
easier to follow than if the lines are set one under
another. A logical numbering will give an order for
setting out the equations in formal work. Figure 3.5
is the digraph for this problem.

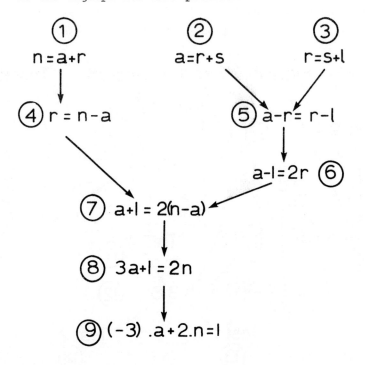

FIGURE 3.5

3.3 Levels

For some purposes logical numberings suffer from the defect that most acyclic digraphs have many logical numberings, and the specification of a particular one may be difficult. On the other hand each digraph has only one assignment of levels to its points. Thus particular features of the levels function of a digraph can be used in testing isomorphism. The levels function is appropriate in some applications also.

> DEFINITION 3.3 The levels function of an acyclic digraph assigns to each point u of the digraph the length of the longest path of which it is the last point.

THEOREM 3.9 If h is the levels function on an acyclic digraph, then h has the properties
    (i)   If u has indegree zero then h(u) = 0
    (ii) Otherwise h(u) is one greater than the maximum of the values of h over the precursors of u;
and h is the only function with both these properties.

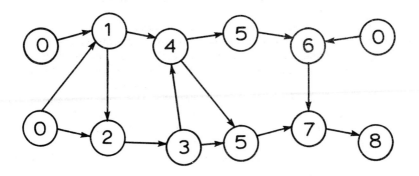

FIGURE 3.6

We omit the PROOF, which is set as an exercise.
    For example of the levels function, Function 3.6 shows an acyclic digraph with the level of each point written in its circle.
    In Figure 2.2 the number part of each course name is one greater than the level assigned to that point by the

function. That digraph has the property that all the
precursors of any point are at the same level, and all the
successors are also at the same level.

3.4 Exercises

1. Prove that the digraph (V,A) with $V = \{a,b,c,d,e,f,g\}$
   and $A = \{(a,b),(a,c),(b,d),(b,e),(c,d),(c,f),(d,g),$
   $(e,g)\}$ is acyclic.

S*2. Consider all logical numberings of the digraph in
   question 1. (There are 29 in all so do not try to
   write them all down). Which points can be number 1?
   Which points can be number 7? What are the earliest
   and last possible numbers that can be applied to d?
   Make a conjecture about the smallest and largest
   numbers that can be given to a point in an acyclic
   digraph, and prove it.

3. Construct a digraph whose points represent the
   garments you are now wearing, and whose arcs represent
   'must be put on before'. Is the digraph acyclic? If
   not, how did you get them on? Find several logical
   numberings. Consider their practicability as dressing
   orders.

4. Construct a digraph on the same lines as Figure 3.4
   for Section 2.2.

5. Find a logical numbering for Figure 2.2.

6-12. For each of the following operations construct an
   appropriate digraph and find a logical numbering.
   Discuss whether your numbering is good or bad from a
   practical viewpoint for the particular problem.
   6: making jam; 7: bottling fruit; 8: making a cup
   of tea or coffee; 9: laying a concrete path;
   10: mending a bicycle puncture; 11: changing a car
   tyre; 12: making breakfast.

13. Construct a digraph for writing a letter, and find a
   logical numbering for it, the points being:
   write letter, post letter, put letter in envelope,
   address envelope, get writing paper, get envelope,
   get pen, get stamp, look up address, seal envelope,
   put stamp on envelope.

S 14. Show that the Rugby score digraph is acyclic.
   (Exercise 1.10.8).

15. Show that the digraph of Exercise 1.10.4 is acyclic.

S 16. Prove that the levels function has properties (i) and
   (ii) as defined in Theorem 3.9.

S 17. Prove the second part of Theorem 3.9, that if f is any
   function which satisfies (i) and (ii) of Theorem 3.9,
   then $f(u) = h(u)$ at every point u of the digraph.

S 18.Obtain the levels function of the digraph in
     Question 1 of these exercises.
  19.Obtain the levels function for the digraph described
     in Question 3 of these exercises.
S*20.Devise an algorithm for calculating the levels function
     from a successor table.
* 21.We can analogously define a function which assigns to
     each point the length of the longest path of which it
     is the first point. State a theorem, corresponding
     to Theorem 3.9, for this function. Prove this theorem
     using Theorem 3.9 and the converse of a digraph.
  22.Construct a digraph showing the stages in eliminating
     v from the equations

$$x = \frac{2}{1+4v^2} \quad ; \quad y = \frac{-4v}{1+4v^2} \quad ,$$

     to obtain $(x-1)^2 + y^2 = 1$.
  23.Construct a digraph whose points represent the sets
     of digraphs:
        digraphs; acyclic digraphs; strong digraphs;
     weakly connected digraphs; symmetric digraphs; complete
     digraphs; antisymmetric digraphs (in which at most one
     of (a,b), (b,a) is an arc for given {a,b}).
        The arcs (A,B) are to represent 'B is contained in
     A'.
     Is the resulting digraph acyclic?

3.5 Essential and inessential arcs

     In Section 3.1 we defined logical numbering in terms
of the numbers applied to the beginning and end of each arc.
In fact many arcs have no effect on the possible
assignments of numbers to their beginning and end points.
We shall describe these arcs as inessential, and simplify
the digraph by deleting them. But first we show that
there is an alternative description of logical numbering
in terms of reachability.

THEOREM 3.10 A logical numbering on a digraph D with p
points is a function n which assigns to each point u of D
an integer n(u) so that:
     (i)   each of the integers 1,2,...,p is assigned to
exactly one point;
     (ii) if v is reachable from u and $v \neq u$, then
$n(u) < n(v)$.

PROOF Except for (ii) this statement is exactly Definition
3.2, so we have to show two things; (a) that (ii) of this

theorem follows from Definition 3.2, and (b) that (ii)
of Definition 3.2 follows from the statements of this
theorem.

(a) Suppose that v is reachable from u, with $v \neq u$.
Then we have a walk consisting of at least one arc from
u to v.
If this walk is $\langle u, w_1, w_2, \ldots, w_m, v \rangle$, then

$$n(u) < n(w_1) < n(w_2) \ldots < n(w_m) < n(v).$$

Hence $n(u) < n(v)$, as was required.

(b) If (u,v) is an arc then v is reachable from u, and
$v \neq u$. Hence by (ii) of this theorem, $n(u) < n(v)$.   □

THEOREM 3.11   If u and v are two points in an acyclic
digraph D, and if for some logical numbering of D,
$n(v) - n(u) = k > 0$, then there are at most 2^{k-1} walks from
u to v.

PROOF   For $0 < i < k$, let w_i be the point such that
$n(w_i) = n(u) + i$. Every point on every walk from u to v
is one of these points, and they must occur in order of
increasing i. As there are $k-1$ values of i, there are
2^{k-1} subsets of $\{w_1, w_2, \ldots, w_{k-1}\}$. Every walk corresponds
to one subset, but not every subset corresponds to a walk
in general; indeed there may be no walks at all. Hence
there are at most 2^{k-1} walks from u to v.   □

The main importance for us is not that the result
gives any sort of estimate of how many walks there are with
given first and last point, but that it establishes the
set of walks to be finite. Some of the network methods of
Chapters 6 and 7 depend on the finite nature of the set of
walks.

We exploit Theorem 3.10 to help us simplify the
digraph before assigning a logical numbering.

DEFINITION 3.4   Let (V,A) be an acyclic digraph, and
let (u,v) be an arc of (V,A). Then (u,v) is
inessential if there is a walk from u to v which does
not traverse (u,v), and essential if there is no such
walk.

THEOREM 3.12   If (u,v) is an inessential arc in an acyclic
digraph (V,A), and B is the set obtained by deleting (u,v)
from A, and if x and y are points in V, then y is
reachable from x in (V,B) if and only if y is reachable
from x in (V,A).

PROOF   If y is reachable from x in (V,B) then there is a

walk whose arcs are in B whose first point is x and last
point y. But all these arcs are also in A, so y is
reachable from x in (V,A).

If y is reachable from x in (V,A), each walk from x to
y will be in (V,B) also unless it contains the arc (u,v).
Every walk containing (u,v) is of the form PQR, where
$Q = \langle u,v \rangle$. But there is a walk W whose first point is u
and last point v, so that there is also a walk PWR with the
same first and last points as PQR. Hence in either case
there is a walk whose first point is x and last point is y
in (V,B). This completes the proof of the converse part. $\square$

Hence deletion of inessential arcs one at a time does
not affect reachability. But it must be observed that the
arc to be removed is to be inessential at the time it is to
be removed and we must consider the possibility that the
status of an arc might change as arcs are removed. Our
next theorem shows that this does not in fact happen.

THEOREM 3.13 Let (V,A) be an acyclic digraph, and let
(u,v) be an inessential arc in (V,A). If B is the set of
arcs obtained from A by deleting (u,v), then an arc (x,y)
is inessential in (V,B) if and only if it is inessential in
(V,A).

PROOF If there is a walk from x to y in (V,B) not using
(x,y), then there is the same walk in (V,A), so that if
(x,y) is inessential in (V,B) it is inessential in (V,A).

Suppose on the other hand that (x,y) is inessential
in (V,A). Then there is a walk from x to y in (V,A)
which does not contain (x,y). If it does not contain (u,v)
it is a walk in (V,B). If it does contain (u,v) write it
in the form PQR, where $Q = \langle u,v \rangle$ and replace Q by W as in
the proof of the previous theorem. There is a possible
snag in that W might contain (x,y), so that PWR could not
be used to show that (x,y) is inessential. But if W does
contain (x,y) we can write W = XYZ, in which $Y = \langle x,y \rangle$, and
then PWR becomes PXYZR, in which both PX and ZR are closed
walks. But (V,A) is acyclic, so the only closed walks are
trivial. If PX is trivial, so is P, and then x = u, and
similarly y = v. But the case (x,y) = (u,v) has been
(implicitly) excluded. So W does not contain (x,y).
Hence PWR is a walk from x to y not containing (x,y) in
(V,B).

Thus in either case (x,y) is inessential in (V,B). $\square$
This theorem implies that we may divide the arcs of (V,A)
into two classes, consisting of essential and inessential
arcs of the digraph respectively. Removing some or all of
the essential arcs does not change the status of the

remaining arcs. This is contrary to the situation in non-acyclic digraphs. The reader may have observed that the definition of an inessential arc does not require the digraph to be acyclic. However the usefulness of such a definition is much reduced without Theorem 3.13, for removal of one inessential arc may convert others from inessential to essential, and the following definition is impossible for non-acyclic digraphs.

> DEFINITION 3.5   Let (V,A) be an acyclic digraph, and let E be the set of essential arcs of (V,A). Then (V,E) is called the <u>essential part</u> of (V,A).

The essential part of a digraph is a partial digraph, so the two following general theorems about partial digraphs apply. The proofs are straightforward and are omitted.

<u>THEOREM 3.14</u>   If (V,B) is a partial digraph of a digraph (V,A) and v is reachable from u in (V,B), then v is reachable from u in (V,A).

<u>THEOREM 3.15</u>   If (V,B) is a partial digraph of the acyclic digraph (V,A), then every logical numbering of (V,A) is a logical numbering of (V,B).

But the essential part has special properties so that we may prove:

<u>THEOREM 3.16</u>   If (V,E) is the essential part of the acyclic digraph (V,A) then v is reachable from u in (V,A) if and only if it is reachable in (V,E).

<u>THEOREM 3.17</u>   If (V,E) is the essential part of the acyclic digraph (V,A), then the logical numberings of (V,A) are precisely the logical numberings of (V,E).

These two theorems, whose proofs may be constructed from the earlier theorems of this chapter, show the importance of the essential part. They allow a digraph to be simplified by the removal of all (or some) of its inessential arcs without changing the reachability properties.

Figure 3.7 shows an acyclic digraph, the inessential arcs being shown by broken lines. One logical numbering is shown.

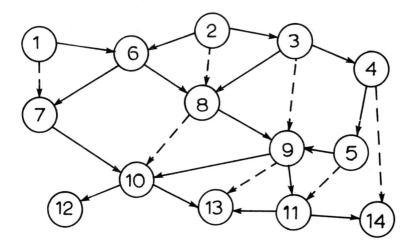

FIGURE 3.7

        In the digraph shown it is not difficult to decide
whether an arc is essential or not.   Some of the rules are:

RULES FOR CLASSIFYING ARCS AS ESSENTIAL OR INESSENTIAL
                    (ARROW DIAGRAM)
(a) If an arc is the only one out of a point, it is
     essential.
(b) If an arc is the only one into a point it is essential.
(c) If there is only one walk from u to v, all its arcs
     are essential.
(d) If (u,v) is an arc and there is a walk from u to v not
     using (u,v), then (u,v) is inessential.
(e) If (u,v) is an arc such that v is assigned the smallest
     number among the successors of u in some logical
     numbering, then (u,v) is essential.
(f) If (u,v) is an arc such that u is assigned the largest
     number among the precursors of v in some logical
     numbering, then (u,v) is essential.

These rules, particularly (c) and (d) may be applied
successively as arcs are deleted.

CLASSIFICATION OF ARCS AS ESSENTIAL OR INESSENTIAL
                    (SUCCESSOR TABLE)
1.   Apply a logical numbering to the digraph and express
     the successor table in terms of it.
2.   Find for each point u the largest number assigned to
     any successor of u. Call this b(u). List b(u) for

D

each point with more than one successor.
3.   If u has only one successor, v, then (u,v) is
     essential.
4.   Follow b(u) by a column in which are to be entered all
     successors of successors of u, and their successors and
     so on so long as the numbers of the points do not
     exceed b(u).
5.   If any successor v of u occurs in the column defined in
     4, (u,v) is inessential: if v does not occur, (u,v)
     is essential.

Point	Successors	b(u)		Essential	In-essential
1	6,7	7	7	(1,6)	(1,7)
2	3,6,8	8	4,5,7,8	(2,3),(2,6)	(2,8)
3	4,8,9	9	5,9	(3,4),(3,8)	(3,9)
4	5,14	14	9,11,14,		
			13,10,12	(4,5)	(4,14)
5	9,11	11	10,11	(5,9)	(5,11)
6	7,8	8	-	(6,7),(6,8)	
7	10		-	(7,10)	
8	9,10	10	10	(8,9)	(8,10)
9	10,11,13	13	12,13	(9,10),(9,11)	(9,13)
10	12,13	13	-	(10,12),(10,13)	
11	13,14	14	-	(11,13),(11,14)	
12					
13					
14					

The bound b(u) ensures that investigation of
reachable points is not carried to unnecessary lengths.
The table above gives the solution for the example shown
in Figure 3.7.
     In this example there are not many inessential arcs.
However, in the remaining sections of this chapter, where
we are using digraphs to represent orderings, there are
very many inessential arcs.

3.6 Partial Orderings (nonlinear ranking)

     In many applications we wish to arrange things in
order of size, importance, age, or other criteria. Very
often the evidence is insufficient for us to decide on a
complete ordering. Sometimes again, the system is not
really linear. There are two approaches to this situation.
The commoner way seems to be to force a linear ranking on
the data by some device. The other is to work within a
system which is sufficiently subtle to be able to handle

the known facts. Acyclic digraphs are such a system, and
in this section we show how they can be used.

A <u>partial</u> <u>ordering</u>, in strict form, may be described
by the properties:

(i)   For any two elements a,b, at most one of the
following is true:
$$a > b, \ a = b, \ b > a.$$
(ii) If $a > b$ and $b > c$, then $a > c$.

These properties are precisely those of reachability
in an acyclic digraph, where $a > b$ is interpreted as a
being reachable from b (or as b being reachable from a)
but $a \neq b$. In this application inessential arcs are
irrelevant and may be suppressed. An alternative approach
is to include further inessential arcs. If we construct a
digraph whose arcs represent the partial ordering, then
condition (ii) implies that there will be an arc (a,b)
whenever b is reachable from a. Such a digraph is termed
<u>transitive</u>, the definition coinciding with the relation
definition. Given any digraph it may be considered as a
partial digraph of a transitive digraph, its <u>transitive</u>
<u>closure</u>, which has the same reachability information and
the same logical numberings. The transitive closure and
the essential part may be looked on as the two extremes of
a family of acyclic digraphs, all yielding the same
reachable sets, the one having all possible inessential
arcs added, the other having them all removed. While the
former may have theoretical advantages, the latter lays
bare the structure.

In this section we develop no new general theory, but
give examples of digraphs representing order situations.

(a)   Subsets.

If we take the subsets of a given set and consider
the strict containment relation (not allowing equality) we
have an example of a partial ordering. We may use
digraphs to show relations between sets when a Venn diagram
is too complicated.

At Canterbury University it is common for first year
students in science and engineering to take four courses
in mathematics: Algebra, Calculus, Mathematical Methods,
Statistics. There are 16 possible subsets of the set of
these four courses. In 1975 the 245 students who took all
four courses passed in 13 of these 16 combinations. But
seven of the combinations accounted for all but eight
students. These seven were as shown in Figure 3.8, with
the inessential arcs suppressed.

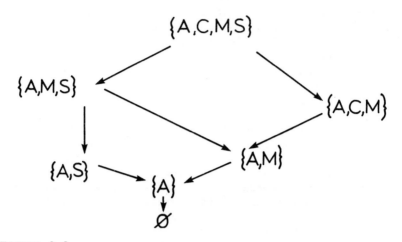

FIGURE 3.8

The figure indicates that generally students found
the Algebra course the easiest but varied considerably,
though systematically, in their response to the other
courses.

(b)   Medal Counts.
At such gatherings as the Olympic Games the newspapers
publish rankings of the nations taking part according to
the numbers of gold, silver and bronze medals won by each.
The ordering is generally linearised by treating gold
medals first;   only if two nations have the same number of
gold medals is the number of silver medals considered, and
only if both gold and silver medals won coincide does the
number of bronze medals influence the ranking.   Thus the
order might be

	Gold	Silver	Bronze
Attica	5	3	2
Sparta	4	1	4
Boeotia	4	0	3
Thessaly	2	5	4
Thrace	2	1	4
Arcadia	2	1	3
Ionia	0	5	0

This ordering assumes that gold medals are worth
immeasurably more than silver, and silver medals
immeasurably more than bronze.   But other views are
possible, such as considering them all of equal value.   The
partial ordering we propose includes both of these as its
extremes.   We assume only the following rules:

(a) A gold medal is worth at least as much as a
     silver medal
(b) A silver medal is worth at least as much as a
     bronze medal
(c) Of medals of any kind, the more the better.

According to these rules, Thessaly beats Arcadia because it
has the same number of gold medals and more of each of
silver and bronze. On the other hand rules (a), (b), (c)
do not allow us to rank Attica and Thessaly, for Attica has
the more gold medals, but Thessaly the higher total so that
the two extreme cases of ranking reverse their order.

Sometimes we need to compute intermediate medal
counts: for instance Boeotia is ahead of Thrace because 4
gold and 3 bronze is better than 2 gold, 2 silver and 3
bronze, which is better than 2 gold, 1 silver and 4
bronze.

The digraph is then Figure 3.9 in which inessential
arcs have again been suppressed.

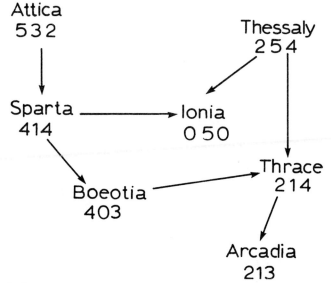

FIGURE 3.9

One problem which occurs in practice particularly
towards the bottom of the list is that different countries
may have identical medal counts. No analysis can separate
such countries: we consider then that we are dealing with
the digraph of medal counts, the condensation of a digraph

in which nations are equivalent if they have the same medal
count.

(c) Geological series.

At any one site, the deeper one goes in the ground,
the older the rocks. At any one place the record will
generally be incomplete. Beds also change their nature
horizontally as material laid at the same time but in a
different place varies in grade and composition. Each site
gives a linear sequence. Identifying beds at different
sites yields a digraph which will be acyclic if the
identifications are correct. The points of the digraph are
the various beds, the arcs indicate that one formation lies
directly above another at some site. In special cases
there may be a spanning walk which will place all the
strata in sequence. More generally the digraph will
suggest that layers at different sites are contemporary.

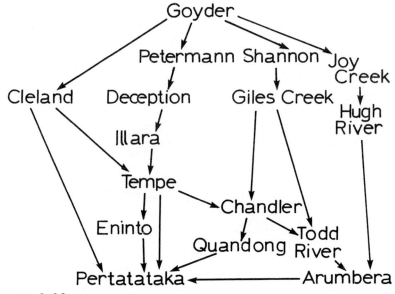

FIGURE 3.10.

In Figure 3.10 we show the interrelation between
strata sampled at several sites in Central Australia.
This digraph includes several inessential arcs.

3.7 Linearisation of Orderings

In the previous section we have seen how to use
digraphs to represent partial orderings, and these will
give a truer picture than forcing a linear ordering on the
set under study. But there are occasions when a linear
ordering is necessary, so it will be useful to develop
techniques for converting an acyclic digraph into a linear
ordering.

Sometimes it will not be necessary to do any violence
to obtain a linear ordering: precisely when there is a
spanning path the points are arranged by the digraph.
These digraphs are discussed under the heading of
unilateral digraphs in Section 4.2. Such digraphs have
exactly one logical numbering, which is the appropriate
linear ordering.

In forcing a linear ordering on an acyclic digraph we
must at least satisfy the condition:

If v is reachable from u, and v $\neq$ u,
then v follows u in the ordering.

Any logical numbering will do as a linear ordering,
and conversely any linear ordering must correspond to some
logical numbering. But there are, except in the case
already mentioned, many logical numberings, and no obvious
way to specify a particular one. We could of course obtain
all logical numberings and then assign to each point the
average of the numbers it received in the ordering, but
this would be very tedious.

The method we give here uses only the numbers of
points in the various reachable sets, so is easy to
compute in the context of the calculations of this and the
previous chapter.

We compute first a <u>ranking coefficient</u>.

Let $r^I(u)$ and $r^O(u)$ be the numbers of points in the
inreach and outreach of u respectively. Both of these
numbers are at least 1 and at most p, where p is the
number of points in the digraph.

We define t(u), the value of the ranking coefficient,
by

$$t(u) = \frac{r^O(u) - r^I(u)}{p - 1}.$$

Then $-1 \leqslant t(u) \leqslant 1.$

To obtain the linearised ordering we take the points
in decreasing order of t(u).

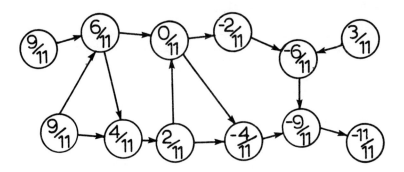

FIGURE 3.11

In Figure 3.11 we repeat Figure 3.6, but with the value of the ranking coefficient inserted for each point. One problem occurs which cannot be avoided in any system which has no arbitrary element is that two points may be awarded the same value of the coefficient. This must occur at least when the digraph has non-trivial automorphisms. (See Section 1.13).

A merit of this ranking coefficient is that it depends only on reachability, so it is unaffected by deletion of inessential arcs.

3.8 Exercises

S 1. Prove that if (u,v) is an arc in an acyclic digraph such that the level of v is one greater than the level of u, then (u,v) is essential. Prove or disprove the conjecture that all arcs where the beginning and end differ in level by more than 1 are inessential.

S 2. Classify the arcs of Figure 3.4 as essential or inessential.

  3. Classify the arcs of Figure 3.6 as essential or inessential.

S 4. Prove Theorem 3.14.

S 5. Prove Theorem 3.15.

  6. Prove Theorem 3.16.

  7. Prove Theorem 3.17.

  8. Find the essential part of Figure 3.10.

  9. How many inessential arcs have been suppressed in Figure 3.8?

10. Construct a successor table for the digraph 'divides
    and is not equal to' on the set $\{1,2,3,6,12,16,24,36,$
    $48,64\}$. Determine which arcs are essential, and
    which inessential by the table method.
11. If two lunar craters overlap, the newer obliterates
    the older in that region. Construct the digraph of
    relative age for the lunar cluster shown in Figure
    3.12. If the craters were formed in periods of
    activity, with no two overlapping craters being
    formed in the same period, what is the minimum number
    of periods of activity?

FIGURE 3.12

12. Find the ranking coefficients for the points of the
    digraph in Exercise 10.
13. Find the ranking coefficients for the lunar craters
    in Figure 3.12.
14. Find the ranking coefficients on the geological
    strata in Figure 3.10.
15. Prove that if two points u,v in a digraph have ranking
    coefficients $t(u) = t(v)$, then neither is reachable
    from the other.

16. Partway through the Commonwealth Games in Christchurch
    in 1974 the leading nations in the medal count were:

	gold	silver	bronze
Canada	8	5	3
Australia	7	6	10
England	3	7	5
New Zealand	3	1	3
Kenya	3	1	2
Northern Ireland	2	0	0
Wales	1	4	0
Jamaica	1	0	0

No other nation had any gold medals.   Construct the
partial ordering.

S 17. An uncompleted test of a mouse's food preferences (see
    Section 4.1) resulted as follows:

Walnut	preferred to millet, wheatgerm, oatmeal, rice, lentils
Wheatgerm	millet, cheese, lentils
Bread	millet, rice, lentils
Oatmeal	millet, rice, lentils
Rice	millet
Cheese	millet, lentils
Lentils	millet

Show that the choices correspond to an acyclic
digraph and calculate the ranking coefficients.

S 18. What happens if you apply the ranking coefficient to a
    digraph which is not acyclic?.

S 19. In Figure 3.10 what is the minimum number of sites
    whose records have been combined?

20. Prove or disprove the conjecture that the sum of the
    ranking coefficients over all points of an acyclic
    digraph is zero.

CHAPTER FOUR

TOURNAMENTS

4.1 Introduction

In popular speech a tournament usually means some kind
of sporting competition. Every tournament in this sense
seems to have its own rules about the number of matches to
be played and the way the winner is chosen. We use the
word in a way which corresponds exactly to what are
frequently called 'round robin tournaments.' In such a
tournament each contestant (individual or team) plays each
other contestant exactly once, and scores, say, a point for
a win and nothing for a loss. The winner is then the
contestant with the most wins. In many sports there are
draws (ties): we shall consider mainly the situations in
which there are no draws.

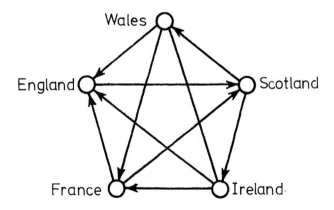

FIGURE 4.1

For example the five 'Nations', England, France,
Ireland, Scotland, Wales, play a Rugby Union Football
tournament each season. In the 1974-5 season it happened
that there were no draws. The results for that season
therefore fit the pattern of mathematical tournaments. We
can show the results by a digraph in which the arc (x,y)
means that team x beat team y. Figure 4.1 shows the
results of the matches.

It is easily checked that this digraph is strong:
there is for example a spanning cycle
        <Wales, Ireland, France, England, Scotland, Wales>.

The application of mathematical tournaments extends to
all areas where the members of some set are compared in
pairs. One field where paired comparison is used is the
general problem of subjective choice. If we wish to study
the food preferences of a laboratory animal we cannot ask
him verbally; we must present him with actual choices.
Presenting many foods at once will confuse both the animal
and the researcher. Instead foods are presented in pairs,
each pair being given together, and it is recorded which is
eaten first. (The problem of deciding which is eaten first
in a practical trial we leave, with relief, to the
experimenter).

As an example of the method we quote the results of an
experiment by H. Gulliksen (Psychometrika, 1956) with 92
college students. They were presented with all pairs of
the meats tongue, pork, lamb, beef and steak. The digraph
Figure 4.2 shows the results of majority choice; for
example given the alternative of tongue and pork the
majority chose pork. We may observe that this digraph is
acyclic: there is also a spanning path
            <steak, beef, lamb, pork, tongue>.

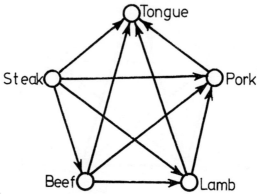

FIGURE 4.2

(It will be implied by Theorem 4.15 that each acyclic
tournament has exactly one spanning path, which visits all
vertices in the order of the logical numbering).
In all such situations we now have the set of things
compared, which forms the point set of the tournament
digraph, and a set of pairs which can be ordered (winner,
loser) and so correspond to the arcs of the tournament
digraph.

DEFINITION 4.1   A tournament is a digraph such that
for any two points u,v exactly one of (u,v), (v,u) is
an arc.
Figure 4.3 shows all the tournament structures on
1,2,3 or 4 points.

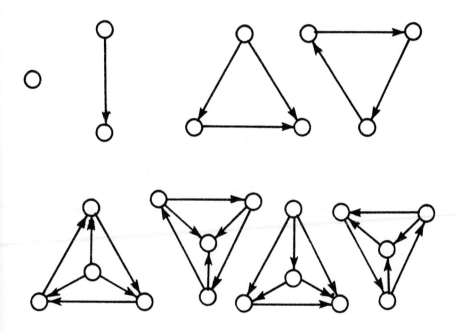

FIGURE 4.3

THEOREM 4.1   For any point u in a tournament with p points,
             id(u) + od(u) = p - 1

PROOF   There is exactly one arc between u and each of the
(p-1) other points.   □

THEOREM 4.2   Any tournament with p points has $\frac{1}{2}p(p-1)$ arcs.

PROOF   Summing the equation of Theorem 4.1 over all points
of the tournament:

$$\sum_{u \in V} (id(u) + od(u)) = p(p-1)$$

But the left hand side is $\sum_{u \in V} id(u) + \sum_{u \in V} od(u)$,

and by Theorem 1.1 each of these sums equals the number of
arcs in the digraph.

Hence   2(number of arcs) = p(p-1).   □

THEOREM 4.3   If u and v are points in a tournament then
either u is reachable from v or v is reachable from u.

PROOF   Either (u,v) or (v,u) is an arc so either $\langle u,v \rangle$ or
$\langle v,u \rangle$ is a walk.   □

Before going on to consider the structure theorems for
tournaments it is convenient to study the structure of all
digraphs with the property that of any two points at least
one is reachable from the other.

4.2   Unilateral Digraphs

DEFINITION 4.2   A digraph D is called unilateral if
given any two points u,v of D, either u is reachable
from v, or v is reachable from u, (or both).

It follows at once that every strong digraph is
unilateral, and Theorem 4.3 could be restated as:

Every tournament is unilateral.

As we have already discovered an acyclic tournament,
it follows that acyclic unilateral digraphs exist. Indeed
the name might be considered unfortunate in that in common
speech 'unilateral' usually implies a strict one-
sidedness which is not the intention of the mathematical
definition.

The unilateral property has immediate consequences for
the inreach and outreach of each point.

THEOREM 4.4   A digraph D = (V,A) is unilateral if and only
if for each point u,

$$R^I(u) \cup R^O(u) = V$$

PROOF   If D is unilateral then for fixed u and any point v,
either v is reachable from u or u is reachable from v.
Hence

$$v \in R^O(u) \text{ or } v \in R^I(u). \text{ Thus } v \in R^I(u) \cup R^O(u).$$

Conversely, if $R^I(u) \cup R^O(u) = V$, any point v belongs to at least one of $R^I(u)$, $R^O(u)$, the former means that u is reachable from v and the second that v is reachable from u.   □

The unilateral property places no condition on the individual components of the digraph, but it does strongly affect the interrelation between them.

THEOREM 4.5 The condensation D^* of a unilateral digraph is unilateral.

PROOF Suppose C_i and C_k are two components of D. Let $u_i \in C_i$ and $u_k \in C_k$. Then in D there is either a walk from u_i to u_k, or a walk from u_k to u_i. Without loss of generality we suppose the former. We have already seen (Exercise 2.8.5) that there is then a walk in D^* from C_i to C_k. Hence D^* is unilateral.   □

THEOREM 4.6 An acyclic unilateral digraph has a spanning path.

PROOF If D is an acyclic digraph it has a logical numbering and we can suppose that its point set $V = \{u_1, u_2, \ldots, u_p\}$ where the suffix is the number given to the point in some logical ordering. Consider two points u_i, u_{i+1} with consecutive suffices. Then as D is unilateral either u_i is reachable from u_{i+1},

or u_{i+1} is reachable from u_i.

The former alternative conflicts with the logical numbering, so u_{i+1} is reachable from u_i. But no walk from u_i to u_{i+1} can contain any third point u_j, for then j would be an integer strictly between i and $i+1$. Hence the only walk is $\langle u_i, u_{i+1} \rangle$, and (u_i, u_{i+1}) is an arc for each i, $i \in \{1, 2, \ldots, p-1\}$. Therefore D possesses a spanning path

$$\langle u_1, u_2, \ldots\ldots, u_{p-1}, u_p \rangle \qquad □$$

These theorems now allow us to describe the component structure of a unilateral digraph. Its condensation is acyclic and unilateral and so has a spanning path

$$\langle C_1, C_2, \ldots, C_k \rangle$$

and can be sketched as in Figure 4.4. There are arcs (C_i, C_{i+1}) for all $i \in \{1, \ldots k\}$ and may be other arcs, all of the form (C_i, C_j) with $i < j$.

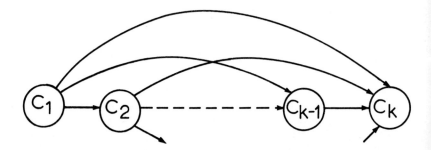

FIGURE 4.4

THEOREM 4.7 A unilateral digraph has a spanning walk.

PROOF If D is an acyclic unilateral digraph with
components $C_1, C_2, \ldots, C_k$, then as the previous theorem shows
there is an arc from C_i to C_{i+1} in D* for each

$i \in \{1, \ldots, k-1\}$.

Hence in D there is a point y_i in C_i and a point x_{i+1} in

C_{i+1} such that $y_i, x_{i+1}) \in A$ for $i \in \{1, 2, \ldots, k-1\}$.

In C we may choose any point as x_1 and in C_k any point as
y_k. (in each case it may be that $x_i = y_i$ even when there

is more than one point in C_i). Within each component C_i
there is a spanning walk with first point at X_i and
last point y_i (from Theorems 2.10 and 2.12). Fitting
together these walks and the arcs between the components
we find a spanning walk

$$\langle x_1, \ldots, y_1, x_2, \ldots, y_2, x_3, \ldots, y_{k-1}, x_k, \ldots, y_k \rangle .$$ □

It happens that the converse of this theorem is also true:

<u>THEOREM 4.8</u> Any digraph with a spanning walk is unilateral.

PROOF Let $\langle u_1, u_2, \ldots, u_k, \ldots, u_p \rangle$
be a spanning walk.
        Let u_i, u_j be two points. Then both appear in the
spanning walk, and the part of the spanning walk between
them is a walk connecting the two points. So one is
reachable from the other.   □
Bringing these two theorems together:

<u>THEOREM 4.9</u> A digraph is unilateral if and only if it has
a spanning walk.
A negative test for unilateral digraphs comes from this.

<u>THEOREM 4.10</u>   In a unilateral digraph there is at most one
point with indegree 0 and at most one point with outdegree
0.
The proof is left as an exercise.
        Because every tournament is unilateral all these
results may be applied to tournaments, as will be done in
the next section. We will also find that the extra
properties of tournaments enable us to strengthen these
results considerably.

4.3 Tournament Structure

        The results of the previous section show at once that
the components of a tournament may be placed in order
$C_1, C_2, \ldots, C_k$ in such a way that if $u \in C_i$ and $v \in C_j$, and if
$i < j$ then $(u,v) \in A$. Hence if we know the components and
the order in which they occur we know the results of each
comparison between objects in different components.
        The special nature of tournaments leads to a theorem
on the possible size of a component:

<u>THEOREM 4.11</u> No component in a tournament consists of
exactly two points.

PROOF In any digraph $\{u,v\}$ is a component if and only if
there is a walk from u to v and a walk from v to u, neither
passing through any other point. This requires that both
(u,v) and (v,u) be arcs. But these two arcs cannot both
appear in a tournament.   □
        The component structure of a tournament is therefore
to a considerable extent described by the cardinalities of
its components and the sequence in which they occur. Each

tournament has a 'component size sequence,' consisting of
the cardinalities of the components in the order in which
they occur in the spanning path of the condensation of the
tournament. The sum of these numbers must be the number
of points in the tournament and 2 cannot occur. These are
the only restrictions. The tournaments in Figure 4.3 have
component cardinality sequences
          1; 1,1; 1,1,1; 3; 1,3; 3,1; 1,1,1,1; and 4
respectively.
     The nature of tournaments also implies that
subdigraphs of a tournament share that property:

THEOREM 4.12 Every subdigraph of a tournament is a
tournament.
The PROOF is immediate and left to the reader.

     DEFINITION 4.3 A subdigraph of a tournament is called
a subtournament.
     It is an immediate consequence of Theorem 4.12 that
every component of a tournament is a subtournament. It is
also strong. Hence we turn to examine strong tournaments.

4.4 Strong Tournaments

     The tournament example shown in Figure 4.1 is strong
as are the first, fourth and last in Figure 4.3. Strong
tournaments are possible for any number of points except 2.
In the example of Figure 4.1 we gave a spanning cycle: we
shall show in this section that every strong tournament has
a spanning cycle. The theorem is in fact more general than
this.

THEOREM 4.13 If T is a strong tournament with p points and
$3 \leqslant k \leqslant p$ then there is a cycle in T which passes through
exactly k points.

PROOF Let u be a point of T. Then the points of T can be
divided into three sets, $\{u\}$, W and X, where W consists of
those points w such that (w,u) is an arc and X of those
points x such that (u,x) is an arc. As T is strong (and
has at least three points by the implications of the
statement of the theorem), neither W nor X is empty. If
there is no arc from any vertex in X to any vertex in W
then no non-trivial closed walk can pass through u, for it
would be of the form
          $\langle$u, x, ..., w, u$\rangle$
and there is no walk from X to W.

Hence T must have points $w_1 \in W$ and $x_1 \in X$ such that (x_1, w_1) is an arc of T. Then

$$<u, x_1, w_1, u>$$

is a cycle of length 3 in T.

Now suppose that we have a cycle

$$<u_1, u_2, \ldots, u_k, u_1>$$

in T, and that there is at least one point of T not in the set $U_k = \{u_1, u_2, \ldots, u_k\}$.

First, there may be a point $v \in V \backslash U_k$ such that for some i, (u_i, v) and (v, u_{i+1}) are both arcs of T (where u_{k+1} is interpreted as u_1). Then

$$<u_1, u_2, \ldots, u_i, v, u_{i+1}, \ldots, u_k u_1>$$

is a cycle through k+1 points.

On the other hand there may not be any such point v. In which case each point not in U_k is either such that all members of U_k are its successors, or all are its precursors. We then divide $V \backslash U_k$ into two sets W_k, X_k : for $w \in W_k$, (w, u_i) is an arc for each

$$i \in \{1, 2, \ldots, k\}, \text{ and for}$$

$x \in X_k$, (u_i, x) is an arc for each

$$i \in \{1, 2, \ldots, k\}.$$

We know that $W_k \cup X_k \neq \emptyset$, for otherwise $U_k = V$, so neither W_k nor X_k can be empty, for that would prevent T from being strong. Further T can be strong only if there is a closed walk through members of both X_k and W_k (and U_k). Since no arc begins in U_k and ends in W_k or begins in X_k and ends in U_k, there must be an arc beginning in X_k and ending in U_k. Let (x_k, w_k) be such an arc. Then (leaving out u_2 arbitrarily),

$$<u_1, x_1, w_1, u_3, \ldots, u_k>$$

is a cycle of length k + 1.
(We do not have to leave out any point at all to make a cycle and certainly have free choice of which one to leave out, but since we are introducing two new points, unless we leave one out there might be gaps in the list of values of k for which cycles of length k exist).

The process of increasing the number of points can be continued until k = p. Hence there must always be a spanning cycle.   □

We may observe that the details of the proof require that in constructing the spanning cycle from the cycle through p - 1 points there must be a point v of the kind required, with (u_i, v) and (v, u_{i+1}) both arcs.

To illustrate this theorem we consider the strong tournament shown in Figure 4.5.

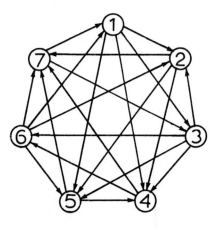

FIGURE 4.5

The proof of the theorem actually demonstrates more than the statement asks: it says that we can choose any point of the tournament and there will be a cycle of length k, for each k, $3 \leqslant k \leqslant p$ through that point. In Figure 4.5 we take the point 1 and find cycles of lengths, 3,4,5,6 and 7 through 1.

To find our initial cycle of length 3, through 1,
$$W = \{6,7\} \text{ and } X = \{2,3,4,5\}.$$
We observe that (2,7) is an arc beginning in X and ending in W, so a cycle of length 3 is
$$<1,2,7,1>$$
With $k = 3$, $u_1 = 1$, $u_2 = 2$, $u_3 = 7$ we notice that (1,3) and (3,2) are both arcs, giving a cycle
$$<1,3,2,7,1>$$
At the next stage there is no $v \in \{4,5,6\}$ such that (1,v) and (v,2) are both arcs, but (3,6), (6,2) are arcs, so
$$<1,3,6,2,7,1>$$
is the next cycle.

Then (3,4) and (4,6) are arcs, and the next cycle is
$$<1,3,4,6,2,7,1>$$
Finally (3,5) and (5,4) are arcs so that
$$<1,3,5,4,6,2,7,1>$$
is the spanning cycle.

This theorem allows us to strengthen Theorem 4.7, that a unilateral digraph has a spanning walk, for the special case of tournaments.

__THEOREM 4.14__ Every tournament has a spanning path.

PROOF Each component of a tournament is a strong
tournament and has a spanning cycle. If component C_i has a
single point this is $\langle x_i \rangle$, where $\{x_i\} = C_i$, otherwise it
has the form $\langle x_i, \ldots, y_i, x_i \rangle$.

Moreover (y_i, x_{i+1}) is an arc for each pair of
consecutive components C_i, C_{i+1}. If we use y_i as an
alternative name for x_i in components with only one point,

$$\langle x_1, \ldots, y_1, x_2, \ldots, y_2, \ldots, x_k, \ldots, y_k \rangle$$

is a spanning path of the tournament.   □

An application of tournaments is in the construction
of sporting tournaments in those sports in which each
competing team has its own 'home' ground. For fairness we
may want each team to play half its matches at home and
half away. We want a way of arranging this.

Because the number of matches each competitor plays is
one less than the number of competitors, this will only be
possible if the number of competitors is odd. Then one
rule is to give each competitor a number (arbitrarily) and
then if i and j are two competitors, the one with the
smaller number is at home if the difference $|i-j|$ is odd
and away if it is even. It is apparent that this gives
each team half its matches at home and half away. When
p = 7 the matches are (home competitor, away competitor):
(1,2), (3,1), (1,4), (5,1), (1,6), (7,1), (2,3), (4,2),
(2,5), (6,2), (2,7), (3,4), (5,3), (3,6), (7,3), (4,5),
(6,4), (4,7), (5,6), (7,5), (6,7).
The analysis described in Section 4.7 will show that these
digraphs are always strong.

4.5 Acyclic Tournaments

At the other extreme from strong tournaments, a
tournament may be acyclic, as the example of Figure 4.2
and the first, second, third and fifth of Figure 4.3 show.
The case is both important, since it corresponds to a
consistent ranking, and easily recognised.

__THEOREM 4.15__ An acyclic tournament has only one logical
numbering and one spanning path. The outdegree of the
point numbered i is p - i, and its indegree is i - 1.

PROOF Every tournament has a spanning path by Theorem 4.14.
Let $\langle v_1, v_2, \ldots, v_p \rangle$
be a spanning path of an cyclic tournament. Then if we let

v_i have number i, the numbering is consistent with the arcs
of the spanning path, and since the digraph is acyclic the
arc between v_i and v_j must be from the one with the
smaller number to the one with the larger in the other
cases also. This fixed the indegree and outdegree also at
the required values. But the indegree and outdegree are
not dependent on the spanning path or the logical numbering.
Hence both spanning path and logical numbering are unique.
□

This relationship between the outdegrees, and between
the indegrees (either directly implies the other) can be
used to prove a tournament to be acyclic.

THEOREM 4.16   Let T be a tournament in p points and suppose
that the outdegrees of the points are
$$p-1, \; p-2, \; \ldots \; 2, \; 1, \; 0$$
then T is acyclic.

PROOF   Let the points be numbered in such a way that v_i has
outdegree p - i, and therefore indegree i - 1. Consider
v_1. It has indegree 0, so for all k the arcs incident with
v_1 are $(v_1, \; v_k)$. In particular (v_1, v_2) is an arc. As v_2
has indegree 1, this is the only arc into v_2. Hence the
arc between v_2 and v_3 is (v_2, v_3). The two arcs into v_3
are (v_1, v_3) and (v_2, v_3), so the arc between v_3 and v_4 is
(v_3, v_4). And so on. This may be converted into a formal
induction proof using the subtournament defined by
$V \setminus \{v_1, v_2, \ldots, v_m\}$.   □

The relationship between the indegrees and outdegrees
can be observed in the examples mentioned at the head of
this section.

4.6   Exercises

S 1. Which of the digraphs in Figure 1.3 are unilateral?
     For those that are, write down a spanning walk.
S 2. Find examples of each of the following:
     (a) A unilateral digraph which has a subdigraph
          which is not unilateral
     (b) A digraph which is not unilateral which has a
          non-trivial subdigraph which is unilateral
     (c) A unilateral digraph which does not have a
          spanning path.
S 3. Prove that the converse of a unilateral digraph is
     unilateral.
* 4. Prove that an acyclic digraph has a unique logical
     numbering if and only if it is unilateral.

5. Prove that every digraph with a unilateral partial digraph is unilateral.

S 6. If D is unilateral digraph and u,v are points of D show that either
$$R^0(u) \subseteq R^0(v) \text{ or } R^0(v) \subseteq R^0(u).$$

S 7. Prove Theorem 4.10.

8. Prove that a subdigraph of a tournament is a tournament.

9. Prove that the condensation of a tournament is a tournament.

10. Prove that the converse of a tournament is a tournament.

11. In a paper (Behaviour 1967) Struhsaker examined the display behaviour of a group of five vervet monkeys, here called a,b,c,d,e. If (x,y) means that "monkey x displayed at least once to monkey y," and the displays recorded were:
    (a,b), (a,d), (b,d), (d,c), (e,a), (e,b),
construct a suitable digraph and show that it is acyclic and unilateral.

S 12. Theorem 4.12 asserts that every subdigraph of a tournament is a tournament. A property P such that if a digraph D has property P then every subdigraph of D also has property, P is called <u>hereditary</u>. What other properties of digraphs are hereditary?

13. A food preference test run on a mouse by paired comparison resulted in foods being preferred the following number of times:
Hollyhock seeds 3, bread 4, cornflakes 2, walnut 5, oatmeal 1, cheese 0.
[Although this is not a genuine experiment, tests we have conducted seem to show quite consistently that mice are not very fond of cheese].
Show that the mouse's choice is consistent. What was the result of the test of hollyhock seeds against cornflakes?

S 14. Show that if a tournament is not acyclic then there is at least one pair of points with the same outdegree.

S 15. Besides deciding home and away matches, the matches must usually be assigned to days, with no competitor playing more than one match on each day. Prove that with p competitors at least p days are required if p is odd. Arrange the matches in the example with 7 competitors into seven sets, to represent the seven days.

16. In the digraph used as the example to Theorem 4.13, find a cycle of each length through the point 5.

4.7 Outdegree Analysis

The previous section has shown us that an acyclic tournament can be recognised from the list of outdegrees (or equivalently from the indegrees). In this section we show that the whole component structure of a general tournament can be obtained from the list of outdegrees.

We shall reduce the process to a mechanical calculation: the reader may well prefer to read only the statements of the theorems and the linking paragraphs of this section until he comes to the description of the table and learn the process in order to see what it gives him, before coming back to fill in the details.

Suppose that the components are $C_1, C_2, \ldots, C_k$ in order, and that the cardinalities of these components are $c_1, c_2, \ldots, c_k$.

THEOREM 4.17: If T is a tournament and u is a point in component C_i of T then

$$c_{i+1} + c_{i+2} \ldots + c_k \leqslant od(u) \leqslant c_i + c_{i+1} + \ldots + c_k - 1$$

$$c_1 + c_2 = \ldots + c_{i-1} \leqslant id(u) \leqslant c_1 + c_2 + \ldots + c_i - 1$$

PROOF Every member of $C_i \cup C_{i+1} \cup \ldots \cup C_k$ is a successor of u. On the other hand at least one member of C_i, namely u itself is not a successor of u. (And if $C_i \neq \{u\}$ there will be another also).

The proof for indegrees is similar.   □

THEOREM 4.18 If u and v are points in a tournament and $u \in C_i$ but $v \in C_j$ with $i < j$ then
$$od(u) > od(v)$$
$$id(u) < id(v)$$

THEOREM 4.19 If u and v are points in a tournament and $od(u) = od(v)$ then u and v belong to the same component.

Suppose now we partition the points of a tournament into two disjoint sets M and N: that is
$$M \cup N = V; \quad M \cap N = \emptyset .$$

Let M and N have m and n points respectively. Consider the sum of the outdegrees of the points in M. This is the number of arcs which have their beginning in M. It is made up of two parts. First, there are the arcs which begin and end in M. There are $\frac{1}{2}m(m-1)$ such arcs. Then there are the arcs which begin in M and end in N. There may be no such arcs or as many as mn, the latter occurring only if every member of M is a precursor of every member of N.

$$\tfrac{1}{2}m(m-1) \leqslant \sum_{u \in M} od(u) \leqslant \tfrac{1}{2}m(m+2n-1)$$

THEOREM 4.20   Let T be a tournament on p points and let V be its point set. If $V = M \cup N$, where $M \cap N = \emptyset$, then $M = C_1 \cup C_2 \cup \ldots C_i$ and $N = C_{i+1} \cup C_{i+2} \cup \ldots \cup C_k$ if and if and only if the sum of the outdegrees of points in M is $\tfrac{1}{2}m(2p-m-1)$, where $C_1, C_2, \ldots, C_k$ are the components in order.

PROOF We saw informally above that the sum of the outdegrees is at most $\tfrac{1}{2}m(m+2n-1)$. But $m + n = p$, so this expression equals $\tfrac{1}{2}m(2p-m-1)$. This bound will be attained precisely when each arc between M and N has its beginning in M and end in N. This will happen when
$$M = C_1 \cup C_2 \cup \ldots \cup C_i$$
and
$$N = C_{i+1} \cup C_{i+2} \cup \ldots \cup C_k$$
For if M and N are as above, every arc between their arcs begin in M and end in N, and if every arc between them is from M to N no component has members in both M and N and then further every component contained in M precedes every component in N.   □

This result shows us that if we collect together sets of points such that their outdegree sums have the right value, this will enable us to break down the tournament into components. Fortunately we do not have to proceed by hit-and-miss methods.

Theorem 4.18 implies that if we list the points in order of decreasing outdegree we group the points in the various components together. It is then sufficient to test the first m members of the list for each m, add their outdegrees and compare the sum with $\tfrac{1}{2}m(2p-m-1)$, which can be computed in advance. In fact we go a little further in order to simplify the computations.

Suppose the points in non-increasing order of outdegree are $u_1, u_2, \ldots, u_p$. Then we have to compare
$$od(u_1) + od(u_2) + \ldots = od(u_m)$$
with $\tfrac{1}{2}m(2p-m-1)$

Now $\tfrac{1}{2}m(2p-m-1) = (p-1) + (p-2) + \ldots + (p-m)$
So u_m is the last point in the list in its component if and only if
$$od(u_1) + od(u_2) + \ldots + od(u_m)$$
$$= (p-1) + (p-2) + \ldots + (p-m)$$
i.e. $[(p-1) - od(u_1)] + [(p-2) - od(u_2)] + \ldots$
$+ [(p-m) - od(u_m)] = 0$

We call the term
$$(p-i) - od(u_i)$$
the <u>deficit</u> of u_i and write it as
$$def(u_i).$$
Hence the condition for u_n to be the last point in its component is that
$$cum(u_m) = def(u_1) + def(u_2) + \ldots + def(u_m) = 0.$$
where $cum(u_m)$ is called the <u>cumulative deficit</u> at u_m. We Observe from the equation that
$$cum(u_m) = cum(u_{m-1}) + def(u_m).$$
We construct a table to facilitate the calculation.

TOURNAMENT WINS ANALYSIS

1. Construct a table with five columns, headed: point, od, p - m, def, cum.
2. In column 'point' list the points in non-increasing order of outdegree.
3. In column 'od' list the corresponding outdegrees.
4. Let p be the number of points in the tournament. In column 'p - m' write p - 1 in the first row, decreasing by 1 at each row to 0 in the bottom row.
5. In each row subtract the entry under 'od' from the entry under 'p - m' and enter the result in the column 'def'.
6. Copy the first entry in column 'def' into column 'cum'. To obtain the remaining entries in the 'cum' column add the entry to the left to the entry above.
7. Rule across the table under every 0 in the 'cum' column. As a check, the final entry must be 0, and none can be negative.
8. The lines partition the points into the components.
9. End.

As an example we analyse the tournament of Figure 4.6.

Point	od	p-m	def	cum
a	6	7	1	1
b	6	6	0	1
c	6	5	-1	0
d	4	4	0	0
e	2	3	1	1
f	2	2	0	1
g	1	1	0	1
h	1	0	-1	0

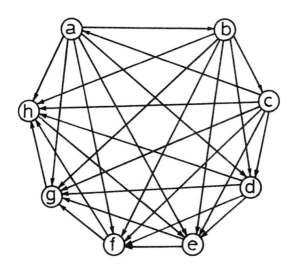

FIGURE 4.6

The components are therefore
{a,b,c}, {d}, {e,f,g,h}.
The reader should also find the components by component
analysis (Section 2.3) and so see that the result is the
same.

From the list of components we can reconstruct most of
the individual arcs. To be precise we know the direction
of all arcs whose beginning and end are in different
components. If we are given a little more information we
can reconstruct the remaining arcs by a chain of reasoning.
In this case, for example, suppose that we are given the
components, outdegrees and the facts that (a,b) and (h,e)
are arcs. Within the component {a,b,c} each point is the
beginning of one arc and the end of one arc. If (a,b)
is an arc, therefore, so are (b,c),and (c,a). Within
{e,f,g,h} the argument is a little more complex. As (h,e)
is an arc and h has outdegree 1, the other arcs at h are
(f,h) and (g,h). As e is the end of one arc within the
subtournament defined by {e,f,g,h}, (e,f) and (e,g) are
arcs. This leaves the arc between f and g, which must be
(f,g) to balance the outdegrees.

This kind of argument can be used generally. Each
component must be treated separately, and the number of
arcs which need to be given before a complete solution can
be obtained rises generally with the number of points in

the component.

4.8 Tournaments with ties

If, as is the case in many applications, we wish to
allow draws, or some more subtle relationship between
points than can be represented by including just one of the
arcs (u,v), (v,u), we may proceed as follows:

We first assign to each pair of points u, v two numbers
numbers dom(u,v), dom(v,u). (We can think of dom(u,v) as
the dominance of u over v), subject to

$$0 \leqslant dom(u,v) \leqslant 1$$

and $dom(u,v) + dom(v,u) = 1.$

We then construct the digraph by including arcs (u,v)
whenever dom(u,v) $\neq$ 0. Thus we have just one arc between u
and v if the dominances are 0 and 1 and both arcs
otherwise.

This approach is useful in various sports where ties
can occur and also each pair of teams meets twice, scoring
two points for a win, one for a tie and none for a loss.
Then four points must be distributed between each pair
teams over their two matches. We obtain dominance by
dividing the points by 4.

This approach shares some, but not all, the
properties of tournaments. The digraph is still
unilateral, but components with two points are now possible.

The component calculation can be carried out as before,
the outdegree being replaced by the sum of the dominances
but all other quantities and calculations remaining the
same.

As an example we analyse the international field
hockey tournament played at Christchurch, New Zealand in
1974. Each of the seven teams played each of the others
once, but draws occurred. Thus we assign dominances 1 for
win, $\frac{1}{2}$ for draw and 0 for loss. We list in order of the
sum of the dominances.

Team	dom	p-m	def	cum
Pakistan	6	6	0	0
Australia	$4\frac{1}{2}$	5	$\frac{1}{2}$	$\frac{1}{2}$
New Zealand	4	4	0	$\frac{1}{2}$
Netherlands	$3\frac{1}{2}$	3	$-\frac{1}{2}$	0
Poland	$1\frac{1}{2}$	2	$\frac{1}{2}$	$\frac{1}{2}$
Canada	1	1	0	$\frac{1}{2}$
Malaysia	$\frac{1}{2}$	0	$-\frac{1}{2}$	0

Thus Pakistan forms a component on its own and there are
two components of three each. On this occasion we can
also deduce some of the results of matches between other
teams, provided, as is often the case, we know not just the
dominance totals but also the numbers of wins and draws
separately. On this occasion it is sufficient to know that
there were only two drawn matches. As these must be
between teams in the same component, the drawn matches must
have been Australia v Netherlands and Poland v Malaysia.
Then Malaysia lost all of its other matches, particularly
that with Canada which accounts for all of Canada's
dominance total. Hence Poland beat Canada. A similar
analysis shows that New Zealand beat Netherlands and
Australia beat New Zealand.

4.9 Exercises

S 1. Suppose a tournament has nine points. If there are
     two components, and exactly half the arcs link
     points on the same component, find the sizes of the
     component.

S 2. The first round of the World Softball Championships
     at Lower Hutt, New Zealand in 1976 resulted in the
     following wins: New Zealand 5, Canada 5, United
     States 5, Japan 3, Taiwan 2, Guam 1, South Africa 0.
     Find the components, and given that Canada beat the
     United States find the results of all matches.

  3. Stephen was asked to choose his preferred sandwich
     filling by comparing each pair in turn. The number
     of times each was preferred was as follows:
     Ham 11, peanut butter 9, honey 9, vegemite 8, grated
     cheese 7, potato 6, marmalade 5, lettuce 5, kernel
     corn 3, cucumber 2, cream corn 1, tomato 0.
     Find the components of the tournament.

S 4. A tournament resulted as follows:

Competitor	A	B	C	D	E	F	G	H	I
Wins	8	6	6	6	4	2	2	1	1

     Find the components. Of how many matches can you not
     now give the result?

  5. Nine children played a tournament. The number of wins
     scored by each was
     Ham 6, Bev 5, Susan 5, Tony 5, Kim 4, Peewee 3,
     Rohan 3, Barry 3, Tom 2.
     Show that the digraph is strong.

6. A tournament resulted in the following -

Contestant	Beat
A	E, G, I
B	A, F, G, H, I
C	A, B, E, F, G
D	A, B, C, F
E	B, D, F
F	A, G, H
G	D, E
H	A, C, D, E, G, I
I	C, D, E, F, G

Show that the tournament is strong. Find cycles of every length from 3 to 9.

7. A tournament ended with the following wins table

Player	A	B	C	D	E	F	G	H	I	J	K	L	M	N	O	P	Q
Wins	15	15	15	13	12	10	10	9	8	8	5	5	4	4	2	1	0

Find the components of the digraph. Find the result of each of the following matches, if it is possible to do so: (A,D), (F,I), (G,K), (P,Q).

S 8. In a sports competition among seven teams, each pair of teams met once. Two points were awarded for a win and one point each for a tie. The first three teams in the final table were:
Apes: 11 points; Bears: 10 points; Cats: 9 points.
What were the results of the matches between them?
There are two solutions.

CHAPTER FIVE

ROOTED TREES

5.1 Introduction

The purpose of this chapter is to introduce a special
class of digraphs called rooted trees. A rooted tree
is a digraph with a unique proper source, called its root,
which has the property that there exists a unique walk from
the root to each other point. Some examples of rooted
trees are shown in Figure 5.1. The roots are circled.
These types of digraphs are particularly useful for
solving problems where the solution is obtained by
considering various cases of combinations of elements. The
rooted tree is used to ensure that all relevant cases are
considered. This is best explained by means of examples,
some of which are given in the next section.

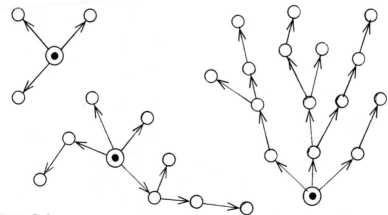

FIGURE 5.1 Some rooted trees

We begin the study by demonstrating some properties.

THEOREM 5.1   Any rooted tree is acyclic.

PROOF (By contradiction). Suppose that a rooted tree, R contains non-trivial closed walk C. Let p be a point on this closed walk. Let W be the unique walk from the root to p. Then W followed by C is a different walk from the root to p, contradicting the assumption of uniqueness.   □

THEOREM 5.2   In any rooted tree there is no walk to the root from any other point.

PROOF (By contradiction). Suppose that a rooted tree, R contains a walk W from a point p to the root r. Now from the definition there must exist a walk X from r to p. Then the walk W followed by X is a non-trivial walk from p to p. This contradicts Theorem 5.1.   □

THEOREM 5.3   In any rooted tree there is exactly one arc into each point other than the root.

PROOF (By contradiction). Suppose there exist two arcs (u,w) and (v,w) into a point w which is not the root in a rooted tree, R. Now by assumption there is a walk W from the root to u and a walk X from the root to v. Then W followed by $<$u,w$>$ and also X followed by $<$v,w$>$ are different walks from the root to w. This contradicts the assumption of the uniqueness of walks.
     If either u or v is the root, say u, then both $<$u,w$>$ and X followed by $<$v,w$>$ are still different walks from the root to w. This completes the proof.   □
     The converse of Theorem 5.3 can also be proven:

THEOREM 5.4   An acyclic digraph which has exactly one point of indegree zero and all other points of indegree one is a rooted tree.
     The proof is left as an exercise.

5.2   Applications of rooted trees

     Consider the digraph in Figure 5.2 in which the numbers represent arc lengths and the points are labelled with letters. If you were asked to find the lengths of all paths from α to ω how would you go about it? Obviously some sort of systematic approach which methodically records all meaningful combinations of arcs and their lengths is needed. Rooted trees can help us keep track of the right combinations.

Let us begin to enumerate the possibilities.
Beginning at α we may first visit b (length 3), or c
(length 3), or d (length 4). If we have gone from α to b
we may next visit e (total length 6), or c (total length 4).
And so on. A rooted tree showing all possibilities is
given in Figure 5.3.

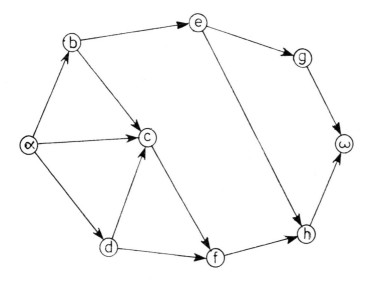

FIGURE 5.2 A digraph for the path enumeration problem.

A rooted tree showing all possibilities is given in
Figure 5.3. The root signifies the beginning of all paths.
The symbols by each point show the label of the point of
the original digraph and the total length from α along the
path to that point. It must be emphasised that as the
same point of the original network can be reached by
several paths, its name will appear as the label of
several points in the tree. Thus 'c' appears three times;
representing the three paths from α to c, namely $\langle\alpha,b,c\rangle$,
$\langle\alpha,c\rangle$ and $\langle\alpha,d,c\rangle$. The label is thus a shorthand for a
path name; the path itself can be read from tracing
through the tree. The completed rooted tree shows us that
the longest path is $\langle\alpha,b,e,h,\omega\rangle$ and the shortest is
$\langle\alpha,c,f,h,\omega\rangle$, and also that there are six paths altogether.

E

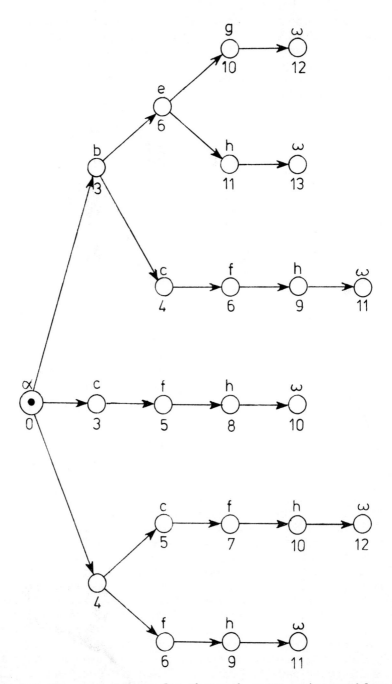

FIGURE 5.3 The rooted tree for the path enumeration problem

   (In Section 5.4 we improve this tree approach to
shortest paths and in Section 6.2 give a completely
different method).
   As a second example consider the dilemma of a market
gardener who can plant only one crop at a time. Further,
each crop can be grown only in a limited range of time
because of the effects of the seasons. Each crop must
occupy the maximum duration of its range in order to be
profitable. The crops, their time ranges, and profits are
shown below:

| | Time range: | | |
Crop	from week	to week	profit ($)
Asparagus (A)	1	17	400
Beans (B)	6	19	450
Carrots (C)	18	27	350
Dutch Spinach (D)	21	32	450
Eggplant (E)	21	47	800
French Radish (F)	28	47	750
Gherkins (G)	28	37	350
Herbs (H)	33	47	500
Iceland Poppies (I)	38	49	350

The problem is to determine which sequence of crops
maximizes the total profit. We can simplify the rooted
tree by observing that if C or D or E ... etc. is the first
crop the sequence does not yield the maximum profit. This
is because any of these later crops can always be
preceded by either A or B, yielding a higher profit. If
we wished to keep the number of possibilities to be
investigated to an absolute minimum we would make use of
facts like "if A is followed by F,G,H or I there is time to
grow C in between", and so on. The final rooted tree is
shown in Figure 5.4. Thus the sequence: asparagus,
carrots and french radish is most profitable at $1,500.

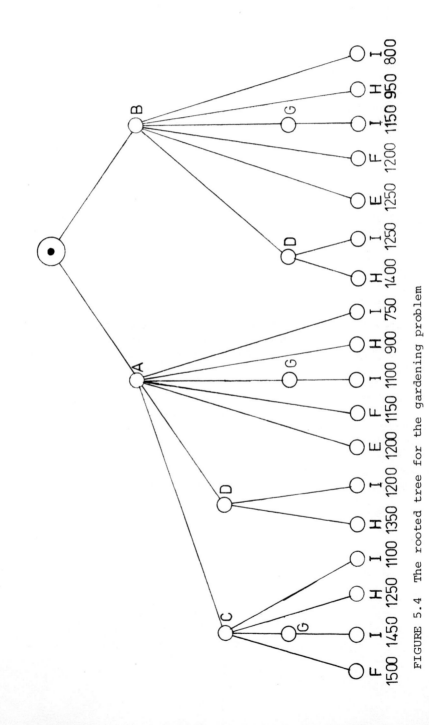

FIGURE 5.4   The rooted tree for the gardening problem

5.3 Exercises

S 1. In New Zealand there are coins valued at 1,2,5,10,20
   and 50 cents. Construct a tree to find all ways of
   making a dollar (100 cents) in at most five coins.

2. In a typical German family tree only the male
   descendants of a man are recorded. The arc (x,y)
   represents the relationship: "x is the father of y".
   Show that the digraph so obtained is a rooted tree.

3. Reduce the rooted tree of the market gardening
   example of Section 5.2 as far as possible by removing
   all sequences which leave a gap large enough to grow
   another crop.

S 4. Four men: Alan (A), Bill (B), Charles (C) and Dave (D)
   dine every night at a circular table with their
   respective wives: Ethel (E), Frances (F), Gladys (G),
   and Hester (H). Each person must have a person of the
   opposite sex sitting on either side. The husbands
   flatly refuse to sit next to their wives. Frances and
   Alan are a little more friendly with each other than
   their spouses realise and always sit together. Solve
   the problem of how many nights they can all dine
   together without a repetition of seating adjacencies
   by constructing an appropriate rooted tree.

5. The Pythagorean philosophers divided their
   mathematics (which did not have the same bounds as
   ours) into two branches, the discrete and the
   continued. The discrete was further subdivided into
   the absolute (number) and the relative (music). The
   continued was divided into the stable (geometry) and
   the moving (astronomy). Draw a rooted tree to depict
   this classification.

S 6. A chemical compound is known to have the formula
   $Na_u H_v S_w O_x$ (so that each molucule contains u atoms
   of sodium, v atoms of hydrogen, w atoms of sulphur
   and x atoms of oxygen). Its molecular weight is
   known to be 120, so that
   $$23u + v + 32w + 16x = 120,$$
   and each of u,v,w,x is a positive integer.
   Construct a rooted tree to obtain all formulae
   consistent with the information.

7. White (Ecology, 1973) records an experiment with
   cicada eggs. The eggs are laid in twigs. If the
   twigs are too small the eggs do not develop. If the
   twig is large the egg may still fail to develop. If
   it does not develop it may be pushed out of the twig
   or stay in it. If it is pushed out it will fall,

either into a container, in which case it will be
preserved, or onto the ground, in which case it will
be lost. If the egg develops the insect eventually
falls to the ground. Again it may either fall into
the container or outside it. Construct a rooted tree
showing the possible fates of an egg.

S 8. A new nation wishes to design a national flag. There
will be three horizontal stripes of colour. Possible
colours are red, green, white, blue and purple. The
following conditions have already been decided on:
(i)   There must be at least one purple stripe.
(ii)   White can only be used for the central stripe.
(iii) Neither red nor blue can be next to purple.
(iv)   The top stripe cannot be green.
(v)   The bottom stripe cannot be blue.
(vi)   Neighbouring stripes can be the same colour, but
       the flag cannot be all one colour.
Find all flags consistent with this description.

S 9. Prove Theorem 5.4.

S 10. Show that if a rooted tree has n+1 points then it has
n arcs.

*S11. Suppose that a rooted tree T has t terminal points
(sinks) and that the longest path in T has length p.
If T has n points, what are the greatest and least
possible values of the product tp?

12. Construct all rooted trees with four or fewer points
(i.e. one member of each isomorphism class).

S 13. Figure 5.5 shows an 'addition' operation on any two
rooted trees; a new rooted tree R+S is formed from R
and S by uniting their roots. Figure 5.6 shows an
operation of forming another rooted tree R^s
("R stalked") from any rooted tree R by introducing a
new root r' joined to the old root r by an arc. Let
T be the trivial rooted tree, consisting of a root
only.
(a)   Draw T, T^s, $T^s + T^s$, $(T^s + T^s)^s$
(b)   Express R of Figure 5.5 in the manner of (a), as
       obtained from T by addition and stalking
       operations repeatedly applied.
(c)   Define R + S and R^s formally, by defining point
       and arc sets for them in terms of the point and
       arc sets of R and S.
(d)   Is it true that for any rooted trees R,S,
           $(R + S)^s = R^s + S^s$?
(e)   Is it true that every rooted tree can be obtained
       from T by addition and stalking operations
       repeatedly applied?

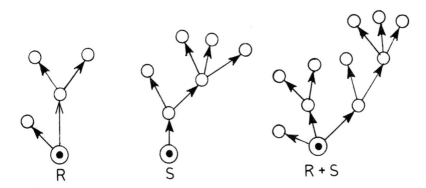

FIGURE 5.5

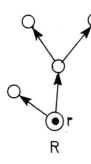

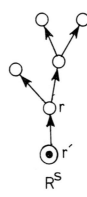

FIGURE 5.6

14. Furnish an original realistic application of rooted trees.

* 15. Let u and v be two points in a rooted tree S. Show that there is a unique point w such that

$$R^I(u) \cap R^I(v) = R^I(w).$$

If we define a binary operation '∧' on the points of R by setting

$$u \wedge v = w;$$

(a) Prove that u ∧ v = u.

(b) Find a digraph description of the relationship between u and v if u ∧ v = u.

(c) Prove that (u ∧ v) ∧ x = u ∧ (v ∧ x) for all points u,v,x of S.

16. In the classification of plants and animals a hierarchy is a collection of sets with the properties:

(a)   the union of all the sets is in the collection
(b)   any two sets in the collection are either
      disjoint or one contained in the other.

The collection {{a,b,c,d,e,f,g}, {a,b,c,d,e},
{f,g}, {a,b}, {c,d,e}, {a}, {b}, {c}, {d}, {e},
{f}, {g}} is a hierarchy. Draw the set
containment digraph (see page 87) and find its
essential part (see page 84). Show that the
essential part is a rooted tree.

* 17. Show that for any hierarchy the essential part of its
      set containment digraph is a rooted tree.
* 18. Show that for any rooted tree the collection of
      outreaches of the various points forms a hierarchy
      under the definition in Exercise 16.

5.4   Branch and Bound Enumeration

Consider once again the path enumeration problem of
Section 5.2. If we had been asked to find the longest path
only instead of all paths we could have still used the same
approach. Once all paths were represented in the final
rooted tree one would check their lengths and find the
longest. Many problems can be solved by this brute force
method of listing all possible solutions and selecting the
best. This listing of solutions is termed enumeration. As
the list of solutions exhausts all possibilities this
approach is called exhaustive or explicit enumeration. It
works well for very small problems, such as the ones we
have looked at so far. However for realistically-sized
problems there can be such an enormous number of
possibilities to be examined that even the fastest
computers would take a prohibitive amount of time.

What is needed for these problems is some way of
reducing the number of possibilities. In other words we
want some quick way of analysis that tells us that the best
solution cannot be among certain large sets of
possibilities. We can discard these sets without having to
look at each possibility individually. Hence the
enumeration is implicit in the sense that we list only
possibilities which have any realistic hope of being the
best. This approach is called implicit enumeration.

One important version of implicit enumeration
involves the construction of rooted trees. This is called
branch and bound enumeration, which is best thought of not
as a single method but as a family of related techniques
united by a common approach.

Before treating a full-scale branch and bound enumeration it is worthwhile looking at the process by which an exhaustive enumeration tree is built up. Consider again the problem considered in section 5.2, the finding of the shortest path from α to ω in the digraph of Figure 5.2

We solved this by building up the tree shown in Figure 5.3. Suppose we built this up in such an order that at some instant we had the partial tree shown in Figure 5.7. Instead of blindly continuing to the completion of the tree let us pause and think.

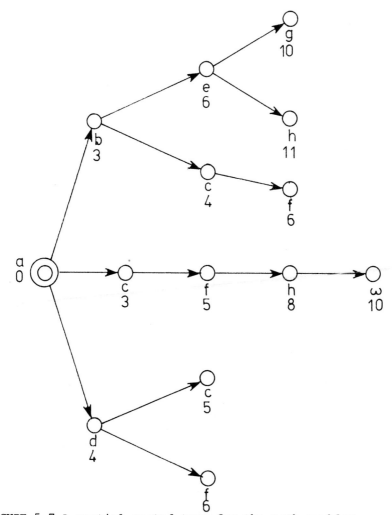

FIGURE 5.7 A partial rooted tree for the path problem.

What we have constructed so far includes one completed
path, $<\alpha,c,f,h,\omega>$ of length 10. We therefore know that at
the worst there is a path of length 10. There may be a
shorter path, but we can observe that there are certain
directions in which it cannot lie. For example, the
pendant point labelled (h,11) represents the partial path
$<\alpha,b,e,h>$ of length 11. However this may be completed to
reach ω, the length will still be greater than 10, so this
will not be the shortest path.

In the same way we may consider the point (g,10) at
the top of the figure. The partial path $<\alpha,b,e,g>$ has
length 10 already and has not reached ω. As there are no
arcs of zero weight, the path will have length greater than
10 by the time it has reached ω, and so not be the shortest
path. Moreover, in these problems we are generally
interested only in finding one best (shortest, cheapest,
most profitable) solution. Thus we can reject (g,10) as
leading to at best another optimal solution.

In the language of branch-and-bound techniques the
function assigning to each path from α to ω its length is
the objective function. We seek the path which minimises
the objective function. (Sometimes branch and bound
techniques seek to maximise, but minimisation is the
general rule). The various paths are called feasible
solutions. These are represented by the pendant points of
the completed tree. The other points of the tree can not
only be thought of as partial paths but also as
representing subsets of the set of feasible solutions.
Thus all paths in this example begin at α, so the root of
the tree represents the set of all feasible solutions.
This set then breaks up into three subsets according to the
second point on the path. These are the points which are
successors to the root. The point labelled (b,3)
represents all paths which begin $<\alpha,b,...>$.

At the point in the investigation shown by Figure 5.7,
we know one feasible solution, $<\alpha,c,f,h,\omega>$. This is
therefore the best feasible solution known at present and
is called the incumbent. The two pendant points at the
top of the figure have been shown not to lead to any
better solution. In the language of branch and bound they,
like the incumbent and any other branch which might have
been pursued to the creation of individual feasible
solutions, or the discovery that there is no way to
complete a certain partial path, are said to have been
fathomed.

The numbers by the points in the tree are the lengths
of corresponding partial paths. In general terms this is

the penalty function. The solution to our problem is a
feasible solution which has least penalty. It is the
essential feature of any penalty function that progressing
along a branch of the rooted tree it never decreases. It
is not necessary for it to increase at every step, as
happens here.

Consider then the present state of the problem. Of
the six pendant points of Figure 5.7, three have been
fathomed, and, apart from the recording of the incumbent,
are of no further interest to us. Were we doing this by
computer and at risk of filling all the store, as often
happens when branch and bound methods are required, we
would take advantage of throwing away or 'pruning' these
parts of the tree.

The three pendant points remaining we might call
active. The next stage of the enumeration is to extend the
partial paths represented by these points step by step
until they are fathomed. This will happen when a penalty
function reaches the value of the incumbent at the time, or
a partial path reaches a point from which it cannot be
extended (which does not happen in this example), or a path
reaches ω and so becomes feasible. In this last case we
must compare its length with that of the incumbent. If it
is the same, or longer, we reject it. If it is shorter the
newly found path becomes the incumbent and the old
incumbent is discarded. The incumbent when all points have
been fathomed is the required shortest path.

If we were applying a full branch and bound method to
this case we would need to specify which of the active
pendant points to branch from next. This we might do by a
rule such as:

Always branch from that active pendant point
whose penalty function value is least. If
this leads to a tie, choose one whose partial
path contains most arcs. If this still leads
to a tie, choose arbitrarily.

In a problem of this size there is of course no point
in such elaboration. At the most we would construct the
tree with an eye open for rejecting anything whose penalty
exceeded that of the incumbent. As the size of problem
increases so it becomes more and more necessary to decrease
the size of the tree actually constructed. We do this
first by adopting a strict rule of search. If this gave
a tree which was still too large we would next have to
look at the penalty function. It might be better to have
a more elaborate penalty function than that here. This
might take more calculating each time, but might save in

that the size of the tree could be much reduced.

We are now in a position to turn to a problem which requires a full-scale branch and bound algorithm.   In this example we are maximising and not minimising.

Consider the problem of Joe the burglar who has entered the living room of a house.   His experienced eye quickly places a monetary value on each of the goods that he thinks is worth taking.   His sack can hold only a limited volume of 100 units.   Which treasures should he steal so as to carry off the most valuable haul?   The data for the problem are given below:

i	item	volume, v_i	value (\$), c_i	ratio v_i/c_i
1	silver cutlery set	30	60	$\frac{1}{2}$
2	silver candle-sticks	50	60	$\frac{5}{6}$
3	stamp album	40	40	1
4	clock	10	10	1
5	four phonograph records	40	20	2
6	framed print	30	10	3
7	bottle of whiskey	10	3	$3\frac{1}{3}$

Note that the items have been ordered in such a way that the the v_i/c_i ratios are in increasing order. This should always be done before the particular branch and bound strategy to be explained is used. Having ordered the items in this way, suppose Joe begins by putting the objects into his sack in this order. After the first two items are in the third brings the total volume up to 120 which is too much for his 100 unit sack. However he notices that the third item is a stamp album. Joe has no scruples about ripping it in half. Let us <u>temporarily</u> make the assumption that part of an object is worth a corresponding fraction of its total value. Then assuming each half of the album is worth the same, Joe stuffs one of the halves, of 20 units in volume, worth \$20 in the sack. His sack is now exactly full. Notice that he placed items in the sack in their order of "value for volume" - the best first. Hence the sack now contains the best possible value: \$60 for the cutlery, \$60 for the candlesticks and \$20 for A to M of grandmother's stamps;   \$140 in all.

Of course, most objects would not retain fractional values when broken into two pieces. To make the problem more practical Joe must give himself the rule: <u>either all of an object is taken, or none of it is</u>;   it cannot be divided.

This makes Joe's problem harder. One thing he does know, and that is that there is no selection of goods which will give him $140, for if so he would have found it by the argument above. If he has time he will put out his pencil and paper and follow a branch and bound algorithm to find his best selection.

He constructs a tree to keep the results of his calculations. He begins by marking the root of the tree. This represents all possible selections of objects with a total volume at most 100. He writes by it "$140", since he knows that no selection has a value greater than that. He proceeds by examining articles in turn, in the order of increasing ratio V_i/C_i.

First he examines article 1, the cutlery set. He must either take it or leave it. So he can divide the set of all selections of items with volume at most 100 into two parts, A and B. (We can't go on writing "all selections of items with volume at most 100", it is too lengthy. So we will simply say "all feasible selections".) Into A go all the feasible selections in which the cutlery is left; into B go the feasible solutions in which it is taken. Between them A and B contain all the feasible solutions, and no feasible solution is in both. As a shorthand way of communicating what is taken and what is left behind we introduce the variable x_i, i=1,2,...,7 where

$$x_i = \begin{array}{ll} 1 & \text{if item i is put in the sack,} \\ 0 & \text{if item i is left behind.} \end{array}$$

Then we can define a solution as a sequence of 0 or 1 values for the variable $x_1, x_2, ..., x_n$. For example

$$(x_1, x_2, ..., x_n) = (1,1,0,0,0,0,0)$$

is the solution in which only items 1 and 2 are taken. We can define the sets A and B as:

A = {feasible solutions in which $x_1 = 0$}
B = {feasible solutions in which $x_1 = 1$}

We represent these sets in the rooted tree by two new points also called A,B as shown in part (a) of Figure 5.8. Now we know that if $x_1 = 1$ we cannot hope for a solution value better than $140. Hence this bound is associated with point B in Figure 5.8. However if $x_1 = 0$ what is the best we can hope for? Joe sets aside item 1 and puts items 2, 3 and 4 in the sack. It is then exactly full and has value

$$\$(60 + 40 + 10) = \$110.$$

Hence this bound is associated with point A. As it is more likely (at our present state of knowledge) that the best

solution lies in set B as it has the higher bound,
investigate first set B, partitioning it into two further
subsets: C and D, which are defined as:
$$C = \{\text{feasible solutions in which } x_1=1 \text{ and } x_2=0\}$$
$$D = \{\text{feasible solutions in which } x_1=1 \text{ and } x_2=1\}.$$
This is represented in the tree in Figure 5.8 in part (b).
This technique of partitioning a set of possible solutions
is known as branching.

We can now find bounds on the best value that solutions
in these sets can hope to attain. (This corresponds to the
penalty function of the earlier example, which was a lower
bound to the value of all solutions in the corresponding
subsets. As we are here maximising what we find is an
upper bound. In a maximising problem it is inappropriate
to speak of a penalty).

We already know that if x_1 = 1, x_2 = 1 then the best
we could hope for cannot be greater than $140. Hence this
bound is associated with point D. What can we hope for if
x_1 = 1 and x_2 = 0? To find out we accept items: 1,3 and 4.
(We could not accept item 2 as x_2 = 0). This yields a
volume of 80 units. The next item, number 5 would bring
the volume up to 120, exceeding the 100 unit limit. Suppose
we accept the fraction of item 5 which brought the volume
up to 100. That is, we accepted half of item 5 with
weight 20. Then, assuming this fraction is worth half the
value ($10) the total value is:
$$\$(60 + 40 + 10) = \$120.$$
Now although we have assumed this process of splitting an
item in two is strictly illegal, we have at least found
the maximum value that we could hope for if we leave out
item 2, namely $120: The half of item 5 may be replaced by
other items, but only at a lower value. (This is why the
items must be in order of increasing v_i/c_i). This is the
bound we associate with point C. Now as point D has the
highest bound of any pendant point we branch from it.
This creates points E and F in part (c) of Figure 5.8. On
examining point F we see that it requires x_1 = 1, x_2 = 1,
x_3 = 1. There are no possible solutions with these values
as these first three items exceed the volume restriction.
Hence this point is pruned from the tree. Calculating a
bound for point E by the method just mentioned yields $135.
As this is the highest value over all active pendant points
we branch from it. Proceeding in this way we build up the
rest of the tree in Figure 5.8. Note that points J and L
are also pruned from the tree. When we get down to the last
last level of points M and N, all items have been
considered. Point M represents a solution which can be

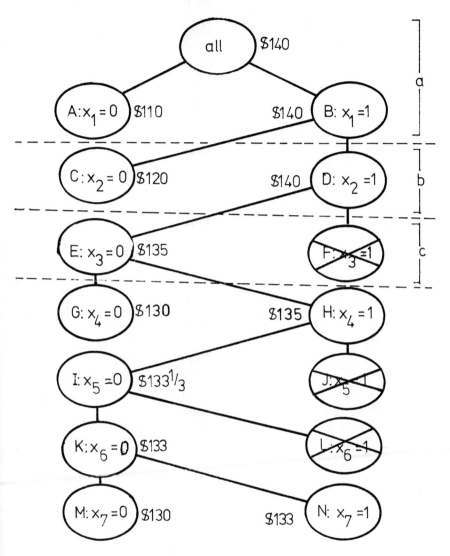

FIGURE 5.8   The branch and bound rooted tree for the
burglar's problem

found by tracing a path from M back to the root:
$x_7=0$, $x_6=0$, $x_5=0$. $x_4=1$, $x_3=0$, $x_2=1$, $x_1=1$. The solution of
point N is the same except that $x_7=1$. Points M and N are
fathomed. So we have found a solution belonging to point N
with a value of $133. All other pendant points have bounds
which are less. Hence there is nothing to be gained by
exploring these other possibilities; solution N which
accepts only items 1,2,4 and 7 is the best. Had there been
an active pendant point with a bound greater than $130 we
would have had to explore it.

Thus it can be seen that branch and bound is a
systematic way of searching through the possible
combinations to find the optimum without blindly listing
them all. It is a very powerful tool useful for solving
many problems associated with digraphs.

5.5 Rooted trees and games

In games like chess, noughts-and-crosses (tick-tack-
toe) or draughts (checkers) where players take turns to make
a move, we can investigate strategies by constructing a
rooted tree showing all the moves. In fact all interesting
games are rather too large for this in practice. Taking
advantage of symmetry and discarding absurd moves it is
just about feasible to construct the tree for noughts-and-
crosses. To describe what happens we take a smaller game
than this.

Suppose there are two players, O and X, and they take
turns, X first, to write their signs in one square in a
row of five. The rule is that O and X may not be written
on neighbouring squares. The first person who cannot move
loses.

Figure 5.9 shows the tree of all possible moves. We
have simplified somewhat by showing only one of any
symmetrical pair of board positions. Having constructed
the completed tree we label the sinks according to which
player wins. We can then work back to the beginning to
show that X must win if he always plays a move shown by a
large arrowhead.

The argument is as follows: Consider a position such
as that marked 'F'. It is X's turn. If he takes the end
square, O will win. If he takes the middle square, he
will win. So he chooses the middle square. But position
F will never occur if O plays correctly, for in position C,
O can ensure a win for himself by choosing G rather than F.
Applying this analysis to all positions, a perfectly
played game will always run A - B - E - K or A - B - E - L,

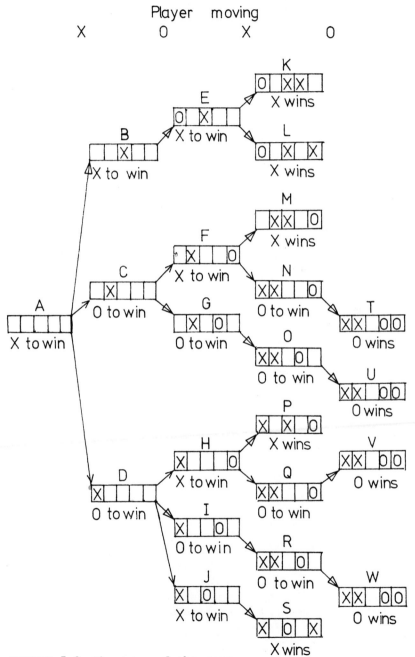

FIGURE 5.9 The tree of the game

either resulting in a win for X.

The reader should investigate games with similar rules
but different boards: a line of a different number of
squares, or a T - or H - shape, or a cross.   The tree will
quickly get much larger as the number of squares is
increased.

The method is not restricted to games with two players,
nor to games in which the players move in rotation.   The
principle to apply in the analysis is that faced with a
choice,   the person whose turn it is will make the move
which most benefits him.   Where he cannot decide, he may
choose any of several alternatives, so each must be marked.
This may make it impossible in some games with more than
two players to decide which will win.

The method may also be used in games where there are
chance moves, when the player who is to move will make the
move which gives him the highest probability of winning.
Again there may be difficulties when there are more than
two players.

5.6   Exercises

S 1. Find the shortest path from α to ω in Figure 6.2.
S 2. The Travelling Salesman Problem.
      A travelling salesman sets off from A to visit B,C,D
      and E in any order and return to A.   He knows the
      distances, as given by the table below: only half the
      table is given because the distances are all
      symmetrical.   Construct a tree, using branch and
      bound ideas, to find the route which gives the shortest
      distance.   Why will there be at least two such routes?

A				
4	B			
7	10	C		
13	11	16	D	
14	17	8	11	E

      (Even a problem as small as this shows that more
      powerful methods are needed.   The Travelling Salesman
      problem is one of the standard notorious problems of
      Operations Research: [Kaufmann] , [Bondy and Murty] .
S 3. The Assignment Problem.
      An employer has to assign five men, A to E, to five
      tasks V to Z, one man to each task.   He knows the
      suitability of each man for each task, which is given

by the table below. He wishes to assign the men so
that the sum of suitabilities over the actual
assignments is a maximum. (For example if the
assignment is
           (A,V), (B,W), (C,X), (D,Y), (E,Z),
the sum of 3 + 4 + 5 + 6 + 1 = 19.

	V	W	X	Y	Z
A	3	2	3	4	9
B	5	4	1	8	3
C	4	1	5	4	2
D	4	7	1	6	5
E	6	7	9	2	1

The problem is best tackled by inverting it: subtract
the table values from 9 and try to <u>minimise</u> the sum of
the new values, which may be considered to be <u>costs</u>.
Consider the various choices for A,B,C,D,E in turn,
always considering next the partial assignment with
least cost.   (There are possible improvements to the
method above, and even better methods are known).

S 4. Solve the following burglar's problem by branch and
bound enumeration:
The volume of the sack = 100.

i	volume, v_i	value ($), c_i
1	10	25
2	50	15
3	60	10
4	20	5
5	10	20
6	30	50
7	40	160

S 5. A game is played on a board like that shown in
Figure 5.10. Player A puts a counter on one circle,
and then B and A moving alternately move the counter
each time one step to a neighbouring circle which has
not been used before. Draw the tree showing all
possible games, and deduce sound strategies (a) when
the last mover wins and (b) when the last mover loses.

S 6. In the example game (Figure 5.9) it is the position on
the board, not the route by which it was reached that
determines which player will win. We may thus turn
Figure 5.9 into an acyclic digraph by identifying the
points (such as N and Q) which represent the same
position reached by different routes. Draw this
digraph and mark on it the analysis.

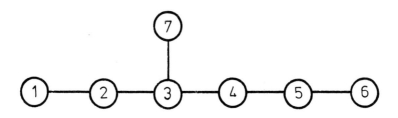

FIGURE 5.10

S 7. Write down the process for analysing a two person
     game on the tree, or the acyclic digraph as defined in
     Exercise 6, as a formal algorithm.
  8. Find, using a branch and bound technique the spanning
     cycle of minimum length for the road network in
     Figure 5.11.

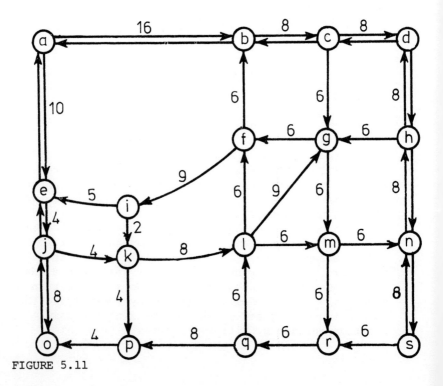

FIGURE 5.11

The figures denote weights of the arcs.   In a two-way
street both arcs have the same length.
[The bulk of the distance is unavoidable and can be
subtracted in various ways so that the penalty function
includes only marginal costs.   It is also worthwhile
to observe that some arcs must appear in every spanning
cycle.   The tree can then be constructed either by
building up a cycle by consecutive arcs, or by
accepting or rejecting arcs in an order which reflects
the key nature of the arc.   One order would be in
increasing order of the sum of the outdegree of the
beginning of the arc and the indegree of its end].

CHAPTER SIX

NETWORKS

6.1 Introduction

Everyone has heard of networks in everyday life.
There are railway, road, telephone and radio networks, and
now even networks of computers in different cities and
even different countries linked together.

In all cases the basic concept is one of communication.
As such the network can be represented by a digraph, in
which the towns, radio stations or computers are
represented by the points and the relation 'can transmit
directly to' is represented by the arcs.

In mathematics the term 'network' is restricted to
certain classes of digraph concerned with some aspect of
communication, and the use of the term varies considerably.
Generally the definition at least includes the requirements
of definition 6.1. We restrict our attention to situations
in which there is a unique proper source α and a unique
proper sink ω, and consider principally the communication
from α to ω. This is not so restrictive as it may appear,
for by various devices many other situations may be
transformed into this convenient form.

Often there is a number associated with each arc,
which may be interpreted as a length, either of time or of
distance. In sections 6.2 and 7.4 we deal with such
applications.

Another major application of networks is to the flow
of commodities, for instance oil through a pipeline. Here
each pipe has a certain maximum capacity, and the flow
into each junction point must equal the flow out. The
problem here is to adjust the flow in particular pipes so

that the flow through the system as a whole is maximised.
Such a problem is known as a <u>transport network</u>.   As the
mathematics required to justify the solution methods is
harder than we have assumed elsewhere, and as there are
available treatments in many books (for example [Berge],
[Harary, Norman and Cartwright]), we have not included them
here.

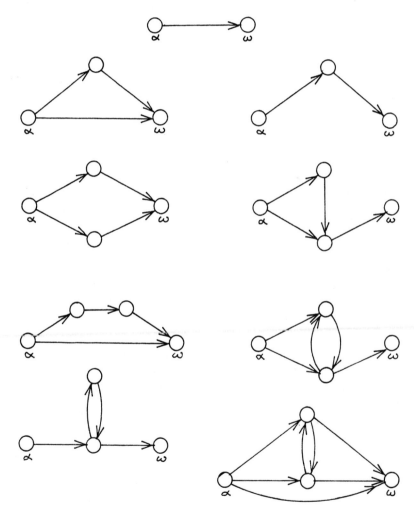

FIGURE 6.1   Some networks on 2,3,4, points

A mathematical network is a special type of digraph.

DEFINITION 6.1 A <u>network</u>, (N,α,ω) is a digraph with
a unique proper source, α and a unique proper sink, ω
such that every point is reachable from α, and ω is
reachable from every point.
Some small networks are shown in Figure 6.1. Some of the
properties of networks are characterised in the following
theorems.

<u>THEOREM 6.1</u> In any network, neither the proper source nor
the proper sink is part of any non-trivial closed walk.

PROOF Consider any point in a non-trivial closed walk.
There is an arc into it, so it is not a source, and an arc
out of it, so it is not a sink.   □

<u>THEOREM 6.2</u> If (N,α,ω) is any network, and N' is the
digraph obtained by adding (ω,α) to the arcs of (N,α,ω),
then any point of N' is reachable from any other.

PROOF Let p,q be any two points of N' (and therefore of
(N,α,ω)). If there is a walk from p to q in (N,α,ω) then
there is still a walk in N'.
    On the other hand whether there is such a walk or not,
there is a walk R from p to ω, and a walk S from α to q.
Hence   R <ω,α> S
is a walk from p to q.
    Hence any point q is reachable from any point p.   □

6.2 <u>Shortest paths</u>

    In the previous section we hinted that the
mathematical study of networks is often used to analyse
real-life problems concerning different types of
communication networks. These problems often have not only
flow along the arcs but also costs associated with the arcs.
These costs may be in terms of time, distance, money or
some other relevant criterion. In the simplest network
problem one is confronted with the task of shipping a given
quantity from the source α to the sink ω at minimal cost,
given that each arc has a shipping cost associated with it.
If the arcs do not have any capacity restrictions this
problem has an easy solution: simply send the whole
shipment by the shortest (least costly) path. This raises
the question of how to find the shortest path. For small
networks one can often work by inspection. However for
larger networks we need a more systematic approach.

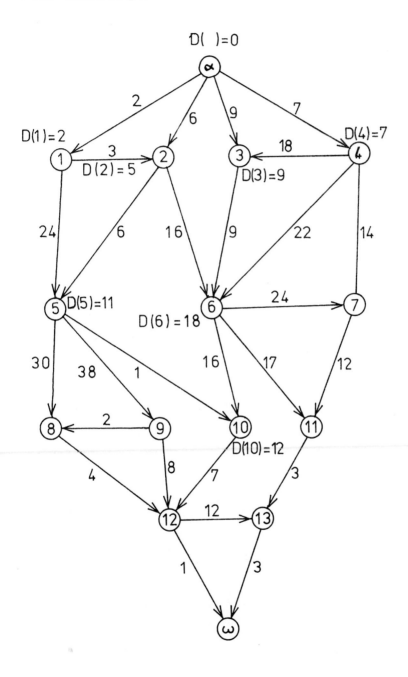

FIGURE 6.2   Finding the shortest α-ω path

A shortest path algorithm

Consider the network in Figure 6.2. How would you go about
finding the shortest path from α to ω?   (The point numbers
are circled, the arc costs are not circled and are next to
the arcs). We achieve this by assigning a label, D to
certain points. The label $D(p_i)$, associated with point p_i
is equal to the length of the shortest path from α to p_i.
When we know $D(\omega)$ all we have to do is to backtrack
through the network to find the actual shortest path.

The shortest path from α to itself is just the trivial
path of length 0 of point α: <α>. Hence
$$D(\alpha) = 0.$$
All the other points are unlabelled as we do not know any
further shortest path lengths yet. We partition the points
into two sets: L, those that are labelled and U those that
are unlabelled. Thus initially
$$L = \{\alpha\}$$
$$U = \{1,2,3,4,5,6,7,8,9,10,11,12,13,\omega\}.$$
Now at each step in the algorithm we find the unlabelled
point which is closest to α. This point can be found as
follows.

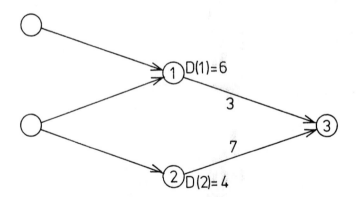

FIGURE 6.3   An illustration of the shortest path algorithm.

Consider Figure 6.3, in which points 1 and 2 have also been
labelled. In this case $1,2 \in L$, $3 \in U$, $D(1) = 6$, $D(2) = 4$
which means that the shortest path from α to 1,(2) is of
length 6,(4). Hence there is a path from α to 3 via 1 of
6+3=9 and a path from α to 3 via 2 of length 4+7=11.
Hence the shortest path we have found to 3 is of length 9.
When using the algorithm we calculate such a shortest path
for all unlabelled points which are successors of labelled

points and choose that unlabelled point and that shortest
path which gives the lowest total. Going back to our first
example, unlabelled points adjacent to a labelled point
(there is only one labelled point, α) are: 1,2,3 and 4.
The shortest path from α to each of these is: 0+2, 0+6, 0+9,
0+7 respectively. The smallest length is that from α to 1
of length 2. Thus point 1 is removed from U and placed in
L. Hence $L = \{\alpha,1\}$,
                $U = A \setminus L$
     and $D(1) = 2$.
On the next iteration we see that points 2,3 and 4 are
adjacent to α and points 2 and 5 are adjacent to 1. Let us
start with point 5. The length to point 1 is 2 and the
(1,5) arc length is 24. Hence the shortest distance from
α to 5 that we have found so far is 26. The distances to
3 and 4 are still 9 and 7 respectively. The distance to 2
via 1 is $D(1)+3 = 2+3 = 5$. The distance via $(\alpha,2)$ is
$D(\alpha)+6 = 0+6 = 6$. Hence we have discovered that it is
shorter to get to 2 via 1 than by going directly. Of all
the shortest distances calculated on this step the one from
1 to 2 is the minimum. Hence 2 is removed from U, placed
in L and labelled:
$$D(2) = 5.$$
The calculations are all recorded in Figure 6.2. We can
now summarize the general step of the algorithm:
     Find
$$\underset{\substack{a \in L \\ b \in U}}{\text{Min}} \{D(a) + c_{ab}\} = D(i) + c_{ij} \ , \qquad (6.1)$$
     where c_{ab} is the length of (a,b).

     Remove   j from U.
     Include   j in L.
     Set $D(j) = D(i) + c_{ij}$.
This is repeated until $\omega \in L$. The rest of the calculations
for the example are:

Iteration:	i	j	D
3	α	4	$D(4) = 7$
4	α	3	$D(3) = 9$
5	2	5	$D(5) = 11$
6	5	10	$D(10) = 12$
7	3	6	$D(6) = 18$
8	6	12	$D(12) = 19$
9	12	ω	$D(\omega) = 20$

Once ω has been labelled the whole process stops. We know
that the shortest path from α to ω is of length 20. To
find the actual path we must unravel the information

contained in the labels. In this unravelling procedure
we can ignore unlabelled points. Working backwards from ω,
we look for the arc which gave ω its label. Recall from
(6.2) that for ω to be labelled there must have been some
point i such that:

$$D(\omega) = D(i) + c_{i\omega}$$

that is

$$D(i) = D(\omega) - c_{i\omega} . \qquad\qquad (6.3)$$

Hence we look for a point adjacent to ω which satisfies
(6.3).

Now
$$D(\omega) = 20,$$
$$D(12) = 19, \quad \text{and}$$
$$c_{12\,\omega} = 1$$

Hence
$$i = 12.$$

We now proceed from i, replacing ω by i in the above step
$$D(12) = 19,$$

$$c_{10\ 12} = 7, \quad \text{and}$$

$$D(1) = 12$$

Hence
$$i = 10.$$

Proceeding in this way we uncover the path from α to ω in
the reverse order: $\langle\omega,12,10,5,2,1,\alpha\rangle$. Hence the shortest
path from α to ω is $\langle\alpha,1,2,5,10,12,\omega\rangle$ with length 20. In
practice for fairly small or manual problems we can remove
the need for these calculations by recording along with
each $D(i)$ the precursor $p(i)$ as we make the forward
calculation. In very large problems this could result in
much more storage being required to keep this information,
most of which may not be used. We would then revert to the
process described above.

6.3   Flowcharts

Flowcharts have been developed largely in connection
with computers. As a computer will do exactly what it is
told to do, and nothing else, it is essential that the
task to be performed is correctly analysed before the
program is written. This is where flowcharts come in. We
do not here apply them to computers but exploit the fact
that they can be used to organise any process consisting of
a number of operations together with certain decision
processes.

A flowchart may be thought of as a network with the
following kinds of points:

1. A single START point with no arcs in and one arc out,

the proper source.
2. A single STOP point with any number of arcs in and no arcs out, the proper sink.
3. A number of INSTRUCTION points, with any number of arcs in and one arc out.
4. A number of DECISION points with any number of arcs in and usually two arcs out. (Computers are normally designed to work on a yes/no basis. However questions with three or more possible answers are feasible for manual operation).

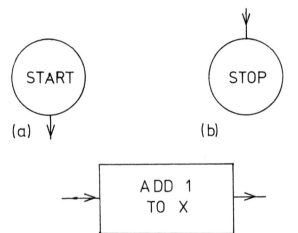

(a)   (b)

(c)

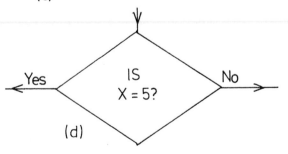

(d)

FIGURE 6.4   The various kinds of points in a flowchart:
             (a) START, (b) STOP, (c) INSTRUCTION,
             (d) DECISION.

The usual shapes of the different kinds of points are shown
in Figure 6.4. In special purpose flowcharts there may be
further coding by shape.

If a flowchart represents some logical task it will
possess an important property:

PROPERTY 6.1   In any logical flowchart, any point p
must belong to a walk which begins at the point
START and ends at the point STOP.

Let us analyse property 6.1. What it is saying is that
p is the endpoint of a walk from START to p and the
beginning point of a walk from p to STOP. The first walk is
necessary or p is irrelevant - there is no way of ever
getting to p. The second walk is necessary or the program
would never stop if p was reached. A flowchart is read by
beginning at START and following through the points,
carrying out instructions as they are met. When a decision
point is reached, the walk taken depends upon the result of
the decision making process.

We shall now illustrate the above ideas by means of an
example. Suppose we wish to find all positive integers less
than a certain number N with a certain property. With some
properties this is a kind of sieve process, successive tests
being made, some resulting in acceptance of the current
number being considered, some in its rejection, some leading
to further tests. The flowchart in Figure 6.5 is of such a
kind. We shall call numbers it accepts "A-numbers".

We will follow through the operation of the flowchart.
Beginning at START we first decide how far we wish to
go. We will set N = 21. We will thus get all A-numbers
less than 21. The number n is the value of the number
currently being considered for acceptance as an A-number or
rejection. We set n = 1. Following the arrow we come next
to a decision point. We ask, "is 1 divisible by 6?" As the
answer is "no" we leave by the arrow directed downwards,
and arrive at the question, "Is 1 divisible by 5?" As the a
answer is "No" again, we leave to the right. A number of
arrows combine and lead upwards to the instruction to add
1 to n: so n becomes 2. We are next asked whether 2 = 20.
As it is not, we leave to the left and return to the point
which asks whether n is divisible by 6. As it is not, we
follow the same route as before. This continues until n = 5
n = 5. Then when we are asked whether n is divisible by 5
we answer "Yes", and leave by the downward path. This leads
to the question whether n+1 (namely 6) is divisible by 3.
As it is, we again leave by the downward route. This brings
us to the information that 5 is an A-number, our first. The

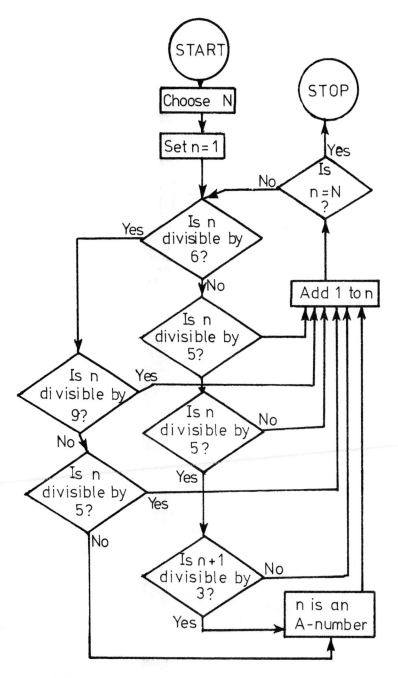

FIGURE 6.5

arrows lead us again round the process.

On the next circuit, with n = 6 we are asked in turn whether 6 is divisible by 6, which it is, and then whether it is divisible by 9, which it is not. We thus find that 6 is our second A-number. As we continue we find that 12 and and 20 are also A-numbers. After n = 20 is declared an A-number, we add 1, and then at the question "Is n = N?" the answer is "Yes", and we proceed to the STOP point.

6.4   Exercises

S 1. Construct all possible acyclic networks with exactly five points and no inessential arcs.

S 2. Construct all possible networks with exactly four points.

3. Suggest an original, real-life application of networks with conservation of flow, arc costs and restrictions on the amount of arc-flow.

S 4. Find the shortest path from α to all points, including ω, in the network of Figure 6.6

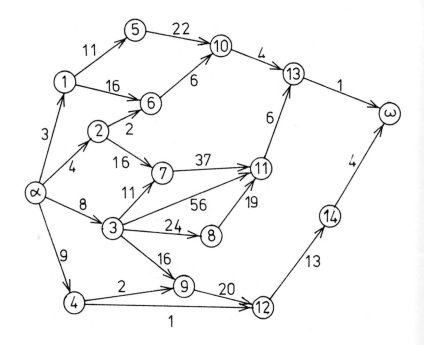

FIGURE 6.6

S 5. Use the flowchart in Figure 6.5 to find all A-numbers
     less than 100.
S 6. Consider the flow chart in Figure 6.7. What does it
     produce in list B?
  7. Construct a flowchart for deciding whether a given
     year is a leap year or not according to the present
     (Gregorian) calendar. (Years ending in '00' are
     leap years only if divisible by 400).
  8. Construct a flowchart for determining which of two
     words should precede the other in a dictionary.
  9. Construct a flowchart for determining the outreach of
     a point in a digraph. (See Section 2.3)
 10. Construct a flowchart for the operation of making a
     cup of tea. Include waiting for the kettle to boil,
     and the options of adding milk and sugar.
 11. Construct a flowchart for constructing the essential
     part of a digraph.
 12. Construct a flowchart for finding the shortest path
     through a network according to the method of Section
     6.3.
S 13. Suppose every arc of an acyclic network has a real
     number, its capacity, associated with it. Let the
     capacity of a path from the source α to the sink ω be
     the minimum of the capacities of the arcs in the path.
     Devise a method for finding a path of maximal capacity
     from α to ω.
 14. Repeat the previous question without the assumption
     that the network is acyclic.
 15. Construct a flowchart governing the operation of
     traffic lights at a cross-roads, where there is a
     pedestrian crossing phase obtainable independently on
     each road on the pushing of a button. You can
     elaborate the system with special turning phases and
     traffic sensors to set the lights green in a
     particular direction if there is no traffic, and so on.
     Since this is a continuous operation, there might not
     be 'START' and 'STOP' boxes.
S 16. Let (N,α,ω) be an acyclic network and let v be a
     point of N. If $p(v)$ is the number of paths from α to
     v, and u_1, u_2, ..., u_r are the precursors of v, show
     that $p(v) = p(u_1) + p(u_2) + ... + p(u_r)$.
     Obtain a similar equation for $q(v)$, the number of
     paths from v to ω.
     Find, in terms of $p(v)$ and $q(v)$ the number of paths
     from α to ω which pass through v.
     If (u,v) is an arc find a similar expression for the
     number of paths from α to ω which contain (u,v).

F

S 17.(Continuation).   Devise algorithms for computing
    p(v) and q(v) as defined in Exercise 16.

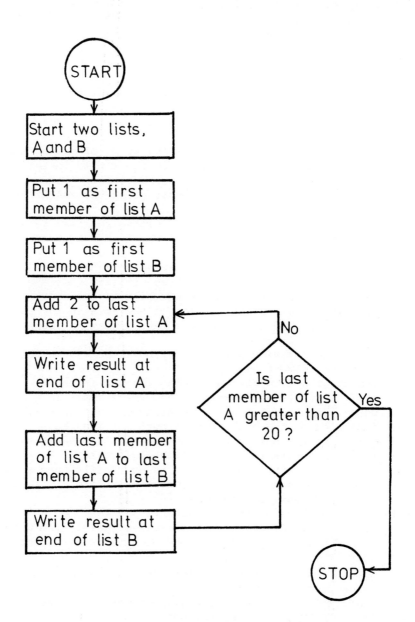

FIGURE 6.7 (see Exercise 7)

CHAPTER SEVEN

ACTIVITY NETWORKS

7.1 Introduction

Many projects in real life, such as the erection of a
building, the preparation for a conference, the design of
a new aircraft or the planning and preparation of an
exploration expedition, can be broken down into a large
number of individual components, which we may call
activities. Often there are constraints between these
activities: one cannot be started until another has
finished. On the other hand, some activities are
completely independent of each other: there is no direct
connection, for example, between ordering the furniture for
a new building and putting in the window glass.
    Thus it is certainly possible to construct digraphs
which indicate the order in which the activities must be
performed: we did this in Section 3.2. In this chapter
we take this analysis further by looking at the time taken,
and try to arrange the activities so that the total
duration of the project is a minimum. In contrast to the
analysis in Chapter 3, when we assumed that only one
activity could be performed at a time, we here assume that
we can do as many activities at a time as we wish - that
are not dependent on each other. Thus we have as many
workers as may be required.
    In any actual project it will turn out that some
activities can take only a short time and can be fitted in
at any point in a long period without affecting the
completion of other tasks. On the other hand some
activities are critical: if they are not done as soon as
it is possible, or if they actually take longer than the

project planner allowed, the completion of the whole
project will be delayed. It is part of the solution
produced in this chapter that the critical activities are
identified. The project planner can then pay special
attention to getting these activities completed on schedule,
and may if necessary divert manpower or equipment from
non-critical activities whose completions may be delayed
without affecting the project as a whole.

The project may therefore be represented by a digraph
in which the points represent the activities, and the arcs
the relationship "X must be completed before Y can begin".
With each point there is also associated a duration time.

As an example of the above ideas consider a small
project such as throwing a dinner party. Let's suppose
this comprises the following tasks:

	Activity:	Precedence:
A.	Choose an evening	–
B.	Plan drinks	A
C.	Buy drinks	B
D.	Plan dinner	A
E.	Buy groceries	D
F.	Cook dinner	C,E
G.	Call friends into dining room	C,F

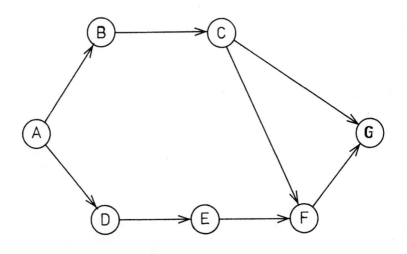

FIGURE 7.1 An activity network

The digraph in Figure 7.1 graphically displays the
relationships between the activities: for instance
activity F cannot begin until tasks C and E are completed.
This is shown in the figure by arrows from C and E to F.
Every digraph which represents a project must be acyclic,
for if there is a closed walk every activity on that walk
must wait for the others before it can begin.   We thus have
an application of the kind discussed under (a) in Section
3.2.
(There are alternative methods of analysis in which the
activities are represented by arcs.   The points are then
called <u>events</u>:   each event may be thought of as the instant
at which all the activities immediately preceding a certain
activity are complete.   This approach will be discussed
further in Section 7.4).
      Once an activity network has been constructed it may
be possible to spot obvious inefficiencies in the over-all
structure of the project.   These can be remedied and the
network layout altered before the analysis begins.   When
the analysis has been completed a schedule can be devised,
giving the time ranges for the commencement and completion
of each activity which will allow the project to be
finished in the shortest possible time.   This will allow
for extra resources to be channelled from activities that
have plenty of leeway into activities that are falling
behind.

7.2   Constructing the network

      In the previous section we glossed over the
difficulties that may come up in actually drawing the
network for a project.   We shall now look into this a
little more deeply.   One must first break the project up
into a number of separate activities - an activity being
defined as any task in the project which takes a positive
amount of time to complete.
      Next each activity is considered in turn.   We ask of
each activity, what activities must be completed before it
can start?   Can other activities be carried out
concurrently with it?
      Once this information has been compiled for all
activities we are ready to draw the network.   First
identify any activity which does not depend upon the
completion of other activities for its start.   A point is
drawn for each such activity near the left hand side of the
page.   If there is more than one such point, they are all
connected to a starting point α further to the left by

arcs directed from this starting point.

Now identify all activities which depend upon the completion of activities corresponding to points already drawn for their start. These points are then drawn to the right of those existing and connected to them by arcs in the direction of the dependent activities. Some of the points just drawn may be inter-dependent and appropriate arcs are drawn in. In no circumstances may a closed walk be created. If this happens, the project has been ill-formulated. The precedence relations must be revised so that the closed walk is broken and the drawing is started again.

Eventually one will have drawn in points for all the activities. There will be at least one which is not required to be completed for the start of any activity. If there is exactly one it is termed the end point and the network has been drawn. If there is more than one, they are all connected by arcs directed toward a new point, termed the end point, ω further to the right.

To illustrate these ideas let us expand the previous example and construct its network. The activites are:

	Activity	Precedence	Time
1	Choose an evening	–	1
2	Choose which friends to invite	–	5
3	Establish which friends can come	1,2	30
4	Store breakables	1,3	20
5	Clean up apartment	4	60
6	Re-arrange furniture	5	10
7	Warn neighbours	3	10
8	Plan drinks	3	6
9	Buy drinks	8	30
10	Make a bar	8	120
11	Plan dinner	3	5
12	Buy groceries	11	60
13	Cook dinner	12	45
14	Call friends into dining room	6,7,9,10,13	1
15	Put a record on stereo	13	1

(We suppose for the purpose of the problem that it is an impromptu party).

We shall now draw the network for the project. There are two activities which do not depend upon other activities for their start: 1 and 2. As there is more than one such point we create a starting point to the left and join it to the points 1 and 2 as shown in Figure 7.2.

What activities depend upon activities 1 and 2? Scanning down the list of precedences we find the answer is activities 3 and 4. (Note that 4 also depends upon 3).

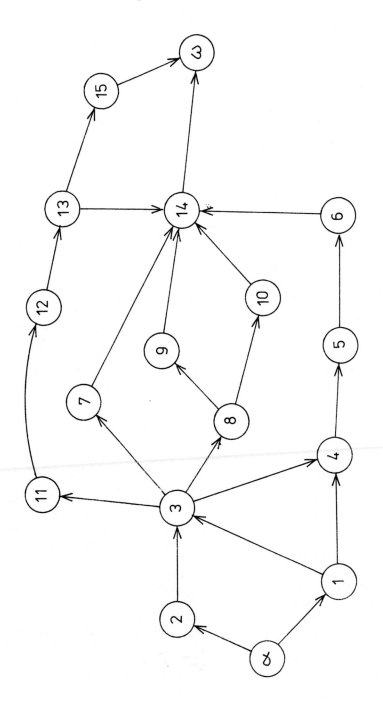

FIGURE 7.2   The construction of an activity network

These points are connected to points 1 and 2 wherever there
is a precedence. An arc is also drawn from 3 to 4 because
of the interdependence of the points just drawn.
Activities depending upon 3 and 4 are 5, 7, 8 and 11.
These are drawn in next together with the appropriate arcs.
This time there is no interdependence between points just
drawn. The next iteration creates points: 6, 9, 10, 12 and
14. Now at this stage we have created point 14 which
depends upon point 13. We cannot draw in the appropriate
arc until 13 is created. This happens in the next
iteration as 13 depends upon 12. Finally 15 is created as
it depends upon 13 and an arc is drawn from 13 to 14.
       Now all activities have been represented. There are
two activities 14 and 15 whose completion is necessary
before any other activity is started. So a further end
point, labelled ω is created to the right and these two
are connected to it. Having checked that all the
precedence relations are reflected in the diagram, the
network is complete.
       Having shown how to construct a diagram of an
activity network we can see that the term network conforms
with our formal definition of a network. Recall that a
network is a digraph with a unique proper source α and a
unique proper sink ω such that every point is reachable
from α and ω is reachable from every point. If we let α
be the starting point and ω be the end point any activity
digraph will be a network.

7.3   The critical path method

       Until now we have not dwelt upon the time it takes to
actually complete each activity. We shall assume that for
each activity the duration time is a given constant. In
many real-life problems that is not true, the times being
random variables. In such cases the times are estimated
and a method called PERT (Programme Evaluation and Review
Technique) is often used to analyse the network. However
such a complication is beyond the scope of this book so we
shall assume constant duration times.
       The purposes of going to the trouble of constructing
a network diagram of a project are twofold. Firstly to
find the shortest possible duration time for the project as
a whole and secondly to devise a schedule which maximizes
the chance that the project will take only this long. We
go about achieving this by associating the duration time of
each activity with its corresponding point in the activity
network. Then as all precedence relations must be obeyed,

the shortest possible time for the whole project
corresponds to the longest start to end path in the network.
(The length of a path is given by the sum of the duration
times of its points).

This path (or paths) is called the critical path and
points on it represent critical activities. Now if any
critical activity is delayed the project completion will
be delayed. Activities which are not critical are termed
noncritical and may be delayed a certain amount of time
without delaying the projection completion. Thus for each
activity we can discover the earliest start time each
activity could possible be begun given that it may depend
upon the completion of other activities for its start. We
define the earliest start time of the starting point
activity to be time zero. We can also discover the latest
finish time each activity can be completed by if the
project is to be completed in the shortest possible time
(on time).

It becomes apparent that a critical activity will have
its earliest start time equal to its latest start time, i.e.
there is no room for slack if the project is to be
finished on time. However in a noncritical activity the
latest start time will be later than the earliest start
time. In this case there is some leeway which we call
float. Two different types of float will be explained later.

So the outcome of a critical path analysis of an
activity network will be a schedule. This will show the
earliest and latest start times and earliest and latest
finish times for each activity. It will highlight critical
activities which will demand special consideration if we
are to get to the church on time. It will also display the
floats for each noncritical activity. Knowledge of these
floats can be used to channel resources from an activity
which has plenty of time up its sleeve to a critical one
for which everything is going wrong and alarm bells are
clanging.

However, noncritical activities become critical if
they are delayed to the point where the float is all used
up. Hence we define the latest start time of an activity
to be the latest time it can be begun if it is not to delay
project completion. Conversely we define the earliest
finish time to be the earliest time that an activity could
possibly be completed given that it depends upon earlier
activities.

We are now in a position to calculate these times for
each activity in a network.

Let ES_i = the earliest start time for activity i,

$\quad$ LF_i = the latest finish time for activity i,

$\quad$ LS_i = the latest start time for activity i,

$\quad$ EF_i = the earliest finish time for activity i,

$\quad$ T_i = the duration of task i.

(If α or ω are dummy activities T_α, T_ω are zero).

It is convenient if the numbers associated with the points form a logical numbering.

$\quad$ We divide up the circle of each point, i as shown in Figure 7.3.

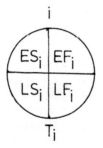

FIGURE 7.3 The labelling of points in a network

$\quad$ The number of the point, i and its duration time, T_i can be written in as soon as the network has been drawn. The other numbers will be filled in by the following process.

$\quad$ ALGORITHM FOR CALCULATING ACTIVITY TIMES

$\quad$ (a) Calculation of earliest times.

$\quad$ The earliest times are linked by the equations:

$\quad$ (i) $\quad$ ES_α = 0, so that we measure time from the beginning of the source activity.

$\quad$ (ii) $\quad$ $EF_i = ES_i + T_i$, for each activity i, since each activity takes time T_i.

$\quad$ (iii) ES_i = max $\{EF_j\}$, where the maximum is taken over all points j with (j,i) an arc of the network. No activity can begin until all the

activities which must precede it have been
finished.

Because the network is acyclic, after ES_α has been set
there must always be a point for which all EF_j for
precursors j are known. If the index numbers i form a
logical numbering the points can be considered in the order
of the numbering. As an example of the calculations
consider Figure 7.4, which is a portion of Figure 7.2
showing the point 14 and its precursors. Supposing the
earliest start times have been calculated for each of the
points the earliest finish of each is found by adding the
duration to the earliest start times. Then the largest of
these earliest finish times is 161 (for point 10). This is
then the earliest start time for activity 14, and since
T_{14} = 1, the earliest finish time for 14 is 162.

The time EF_ω is the duration of the whole project, 162.

(b) Calculation of latest times.

Knowing the earliest completion time of the whole
project, EF_ω, we calculate the latest starting and
finishing times which will allow us to complete the process
by EF_ω.

The equations mirror the earliest time equations.

(i) $LF_\omega = EF_\omega$

(ii) $LS_i = LF_i - T_i$, for each activity i, since each
       activity must begin T_i before it must end.

(iii) $LF_i = \min\{LS_j\}$, for i $\neq$ ω, where the minimum is
       taken over all points j with (i,j) an arc of
       the network.

No activity can be allowed to end after the latest time by
which any successor must be begun.

Because the network is acyclic, after LF_ω has been
set, there must always be a point for which LS_j for all
successors j are known. If the index numbers form a
logical numbering the points can be considered in the
reverse numbering order.

As an example of the calculations consider Figure 7.5,
which is a portion of Figure 7.2. Supposing the latest
finish times for 11,7,8 and 4 to be known, their latest
start times can be found by subtracting the duration of
each activity. The latest finish of point 3 is the least
of these, 35. As 3 has duration 30, its latest start
time is 5.

As a check, the latest start time for α should be
found to be zero:

$$LS_\alpha = 0.$$

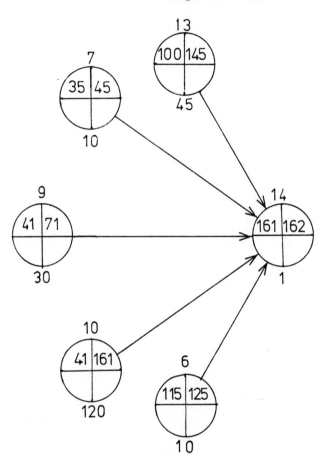

FIGURE 7.4   The calculation of earliest start and finish
              times

        Having calculated the above four activity times for
each activity we are now in a position to identify
critical activities.   Activities, i, which have:

$$ES_i = LS_i$$

are critical.   Such activities will also have

$$EF_i = LF_i \ .$$

        Critical activities form a path or paths from the
starting point to the end point.   Such a path is called a
critical path.   In Figure 7.6 all activity times have been
calculated.   It can be seen that activities: α, 2, 3, 8, 10,
14 and ω are critical and form the critical path.

Having determined the critical path the next task is
to determine _float._ We define _total float,_ TF_i, of an
activity, i to be the difference between the maximum time
available to complete the activity ($LF_i - ES_i$) and its
duration time, T_i.   In symbols:

$$TF_i = LF_i - ES_i - T_i = LS_i - ES_i = LF_i - EF_i$$

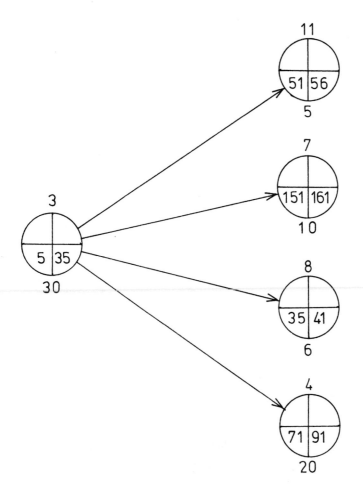

FIGURE 7.5   The calculation of latest start and finish times

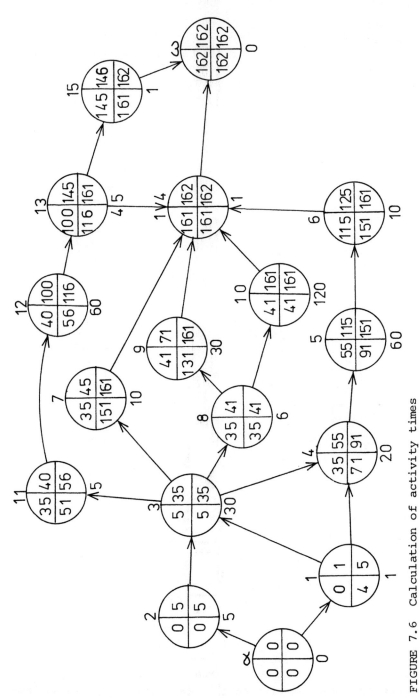

FIGURE 7.6   Calculation of activity times

In Figure 7.6 the latest finish time of activity 11 is time 56 and its earliest start time is 35. Hence the maximum time we have to get 11 done is
$$56 - 35 = 21 \text{ minutes.}$$
Now 11 takes 5 minutes, so we have a total float of 16 minutes. Of course, critical activities will have zero total float.

The second type of float is called <u>free float</u>. It can be motivated using activity 1 in Figure 7.6. Activity 1 has an earliest finish time of 1 minute. Activities 3 and 4 directly follow activity 1. Consider activity 4: it has an earliest start time of 35 minutes. Hence assuming we wish to push each activity through as quickly as possible, neglecting activity 3 for the moment, activity 1 has a float of:
$$35 - 1 = 34 \text{ minutes.}$$
But activity 3 also directly follows 1. Its earliest start time is 5. Thus activity 1 has a float of:
$$5 - 1 = 4 \text{ minutes}$$
We calculate the free float of 1 as the minimum of these, namely 4 minutes. We formally define the <u>free float</u>, FF_i, of activity i as
$$FF_i = \text{Min}_j \{ES_j - EF_i\}$$
where this minimum is taken over all arcs (i,j) in the network. The end point ω has no activities which follow it, and its free float is defined as
$$FF_\omega = 0.$$
Whenever an activity has zero total float it will have zero free float. Examples are the critical activities. However the converse is not true: an activity can have zero free float but positive total float; activity 6 is an example.

We can now complete a schedule for the whole project in the form of a table. This will help a manager make decisions during the running of the project. The schedule for the example is given in the table on the following page.

Activity	Precedence	T_i	ES_i	LS_i	EF_i	LF_i	TF_i	FF_i
α	–	0	0	0	0	0	0	0
1	α	1	0	4	1	5	4	4
2	α	5	0	0	5	5	0	0
3	1,2	30	5	5	35	35	0	0
4	1,3	20	35	71	55	91	36	0
5	4	60	55	91	115	151	36	0
6	5	10	115	151	125	161	36	36
7	3	10	35	151	45	161	116	116
8	3	6	35	35	41	41	0	0
9	8	30	41	131	71	161	90	90
10	8	120	41	41	161	161	0	0
11	3	5	35	51	40	56	16	0
12	11	60	40	56	100	116	16	0
13	12	45	100	116	145	161	16	0
14	6,7,9,10,13	1	161	162	161	162	0	0
15	13	1	145	161	146	162	16	16
ω	14,15	0	162	162	162	162	0	0

7.4 The activity-arc approach

In the previous sections of this chapter points were
used to represent activities and arcs to represent the
precedence relationships between them. There is another
representation in which arcs are used to represent
activities and points are used to represent events that
activities have been completed. Of course these two
representations will give different networks for most
projects. It is possible to deduce one network, given the
other. The activity-arc approach to drawing the network
is the one usually followed in what is commonly called the
Critical Path Method (C.P.M.). However our activity-point
approach of the previous sections is still a C.P.M. method
in the sense that we attempt to detect critical paths.

In fact the activity-point approach enjoys a number of
advantages. There is no need for "dummy" (fictitious)
activities to be introduced in order to ensure that the
precedence relationships are consistent. Dummy activities
often need to be introduced for this reason in the
activity-arc approach. In the activity-point approach
the only dummy activities are α and ω, and
these are not always needed. Also the
activity-point network allows one to make changes to the
network much more easily. Furthermore the four quantities
ES, EF, LS, LF are more directly obtained by the activity-
point approach.

As an example of the activity-arc approach, Figure 7.7
represents the same project as Figure 7.6. We no longer
need the dummy activities α and ω in order to get a network.

The source represents the event 'ready to begin
project'. As activities 1 and 2 have no necessary
preliminaries, they are here represented by arcs out of the
source. They have a common end point 'activities 1 and 2
complete, ready for activity 3'. (Notice that in
Figure 7.6 the arc (1,4) is inessential and can be
removed without affecting the precedence: activity 4
cannot follow activity 1 directly).

As we have drawn Figure 7.7 we do not have a proper
digraph, since these two arcs have the same beginning and
end. As a visual method this gives rise to no problems.

Once activity 3 is complete, activities 4, 7, 8 and 11
can begin. Activities 9 and 10 are in parallel in the
same way as 1 and 2.

Towards the end of the process we have found it
necessary to put in a dummy arc, the broken line β. This
is because 13 is necessary for 14 and 15, but only 13 is

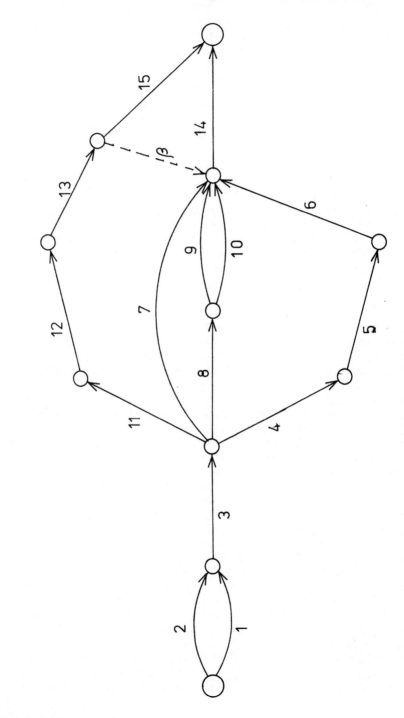

FIGURE 7.7 Activity-arc version of network analysis

necessary for 15 while others are necessary for 14.   To
unite the end of 13 with the beginning of 14 would imply
that 6, 7, 9 and 10 were necessary for 15, which is not the
case.   To leave out β would imply that 13 was not necessary
for 14.
     Activities 14 and 15 are the final activities, so
their arcs meet in the sink.
     The durations now become weights associated with the
arcs.   The critical arcs form <u>critical paths</u> from the
source to the sink.   These paths are the paths for which the
sum of the durations is greatest.   As the network is
acyclic there is only a finite number of paths from source
to sink.   By enumeration, the longest can be found.
However, we desire not only to identify the critical
activities but also the various times of start and finish
of all activities.   There are efficient methods which
provide all this information, along the same lines as those
described for our activity point approach.   These are given
in the texts listed under section (d) of "Further Reading".

7.5   Exercises

     For each of the following projects construct an
activity point network, calculate the values ES_i, LS_i, EF_i,
and LF_i for each point i.   Identify the critical
paths or paths in the network.   What is the earliest time
the project can be completed?   Find TF_i and FF_i for each
point.

	Activity	Precedence	Duration time
1.	1	14,16,13	3
	2	16	7
	3	1,2	9
	4	9,10,3	4
	5	9,10,3	6
	6	4,5	1
	7	–	1
	8	7	8
	9	8	7
	10	8,1,2	4
	11	7	5
	12	7	2
	13	12	3
	14	11,15	16
	15	12	20
	16	17,15	11
	17	11	19

	Activity	Precedence	Duration time
S 2.	1	-	5
	2	-	10
	3	-	8
	4	1	6
	5	1	12
	6	2,4	7
	7	3	4
	8	5,6,7	6
	9	3	10
S 3.	1	-	6
	2	-	4
	3	2	5
	4	2	6
	5	2	4
	6	3	3
	7	4,5	10
	8	7	12
	9	6,8	4

S 4. Consider Exercise 3. Suppose the duration time of
   activity 4 is reduced from 6 to 3. How does this
   affect calculations performed?

  5. Consider the project of painting the exterior of a
   house with two coats of paint. Assume a team of 3 men
   is to carry out the task. Construct a list of about
   10 activities with their duration time. Analyse the
   project using the critical path method.

6-12. For each of the following problems (the same list as
   6-12 in Exercises 3.4), add an estimate of duration
   to each point in the digraph obtained for that
   exercise, and carry out a network analysis:
   6: making jam;   7: bottling fruit;   8: making a cup of
   tea or coffee;   9: laying a concrete path;
   10: mending a bicycle puncture;   11: changing a car
   tyre;   12: making breakfast.

  13. Network analysis assumes that there is sufficient
   manpower to do as many operations as desired
   simultaneously. Examine the solutions you obtained
   in Exercises 6 to 12 to determine the smallest number
   of persons required to carry out the operation in the
   calculated time, bearing in mind that some activities
   may not require any manpower, or only occasional
   supervision.

CHAPTER EIGHT

MATRICES OF DIGRAPHS

8.1 Introduction

A digraph, $D = (V,A)$ is usually defined by specifying
the sets V and A. For instance the digraph in Figure 8.1
can be defined as:
$$V = \{1,2,3,4,5,6,7\}$$
$$A = \{(1,4),\ (1,5),\ (4,5),\ (4,6),\ (5,2),\ (5,6),\ (6,7),$$
$$(7,4)\}$$
This information can be conveniently expressed in the form
of a matrix. In fact there are a number of different
matrices that compactly convey information about a digraph.
These matrices can be manipulated in order to study the
properties of the digraph. This is easily carried out by
computer for relatively large digraphs.

We shall discuss adjacency, incidence, circuit,
fundamental circuit and cut-set matrices. We shall also
present some theorems on the relationship between
properties of digraphs and properties of matrices.
Throughout this chapter we shall consider matrices
concerned with the digraph $D = (V,A)$ which has n points and
m arcs.

8.2 The adjacency matrix

Recall that two points, v_1, $v_2 \in V$ in the digraph
$D = (V,A)$ are adjacent if there exists either of the two
arcs: (v_1,v_2) or $(v_2,v_1) \in A$. (Definition 1.3). Given the
digraph D, its adjacency matrix A (D), is defined by

$$A(D) = (a_{ij}) \qquad i = 1,2,\ldots,n,$$
$$j = 1,2,\ldots,n,$$

where $\quad a_{ij} = \begin{cases} 1 & \text{if } (v_i, v_j) \in A \\ 0 & \text{otherwise.} \end{cases}$

Thus A(D) is a square matrix with as many rows and columns as there are points in D. For each arc in D from, say, point i to point j, there is a 1 in the entry corresponding to the ith row and the jth column. If there is no arc from point i to point j there is a 0 in this entry. The adjacency matrix below corresponds to the digraph defined above and shown in Figure 8.1.

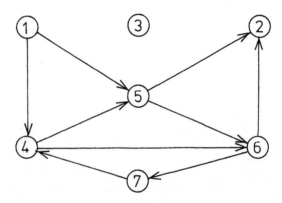

FIGURE 8.1

$$A(D) = \begin{array}{c|ccccccc} & 1 & 2 & 3 & 4 & 5 & 6 & 7 \\ \hline 1 & 0 & 0 & 0 & 1 & 1 & 0 & 0 \\ 2 & 0 & 0 & 0 & 0 & 0 & 0 & 0 \\ 3 & 0 & 0 & 0 & 0 & 0 & 0 & 0 \\ 4 & 0 & 0 & 0 & 0 & 1 & 1 & 0 \\ 5 & 0 & 1 & 0 & 0 & 0 & 1 & 0 \\ 6 & 0 & 1 & 0 & 0 & 0 & 0 & 1 \\ 7 & 0 & 0 & 0 & 1 & 0 & 0 & 0 \end{array}$$

We can deduce a number of properties of any digraph by examining its adjacency matrix. Each column sum equals the <u>indegree</u> of the point corresponding to that column. For example, the sum of column 4 is 2 which equals id(4), the indegree of point 4. Similarly, each row sum equals the <u>outdegree</u> of the corresponding point. For example, the sum of row 6 is 1 which equals od(6). A <u>source</u> can be identified by a column of all zeros, (e.g. point 1), a <u>sink</u> can be identified by a row of all zeros (e.g. point 2).

A proper source can be identified by a column of all zeros which has the property that its equivalent row has at least one 1 (e.g. point 1). A proper sink can be identified by a row of all zeros which has the property that its equivalent column has at least one 1 (e.g. point 2). An isolated point can be detected as a row and corresponding column which both consist of only zeros (e.g. point 3).

Because we have prohibited a digraph from having a loop (an arc of the form (v,v)), the leading diagonal of A(D) consists of all zeros. There exist structures called pseudodigraphs in which it is possible to have multiple arcs from a given point to another point. In this case an entry in the adjacency matrix still represents the number of arcs present, but the entry may be any nonnegative integer.

Of course each entry in A(D) represents the number of walks of length one from the row point to the column point, thus

$$a_{13} = 0$$

indicates that there are no walks of length 1 from 1 to 3, and

$$a_{14} = 1$$

indicates that there is one walk of length 1 ($\langle 1,4 \rangle$) from 1 to 4.

Now let us square A(D) defined on page 170.

$$(A(D))^2 = \begin{bmatrix} 0 & 0 & 0 & 1 & 1 & 0 & 0 \\ 0 & 0 & 0 & 0 & 0 & 0 & 0 \\ 0 & 0 & 0 & 0 & 0 & 0 & 0 \\ 0 & 0 & 0 & 0 & 1 & 1 & 0 \\ 0 & 1 & 0 & 0 & 0 & 1 & 0 \\ 0 & 1 & 0 & 0 & 0 & 0 & 1 \\ 0 & 0 & 0 & 1 & 0 & 0 & 0 \end{bmatrix} \begin{bmatrix} 0 & 0 & 0 & 1 & 1 & 0 & 0 \\ 0 & 0 & 0 & 0 & 0 & 0 & 0 \\ 0 & 0 & 0 & 0 & 0 & 0 & 0 \\ 0 & 0 & 0 & 0 & 1 & 1 & 0 \\ 0 & 1 & 0 & 0 & 0 & 1 & 0 \\ 0 & 1 & 0 & 0 & 0 & 0 & 1 \\ 0 & 0 & 0 & 1 & 0 & 0 & 0 \end{bmatrix}$$

$$= \begin{bmatrix} 0 & 1 & 0 & 0 & 1 & 2 & 0 \\ 0 & 0 & 0 & 0 & 0 & 0 & 0 \\ 0 & 0 & 0 & 0 & 0 & 0 & 0 \\ 0 & 2 & 0 & 0 & 0 & 1 & 1 \\ 0 & 1 & 0 & 0 & 0 & 0 & 1 \\ 0 & 0 & 0 & 1 & 0 & 0 & 0 \\ 0 & 0 & 0 & 0 & 1 & 1 & 0 \end{bmatrix}$$

Let us examine the entry in $(A(D))^2$:

$$a_{16}^2 = 2 \; ;$$

(This symbol $_2$ means the entry in the first row and sixth column of A^2, not the square of the entry in the first row and sixth column of A.)

Now this entry equals the scalar product of row 1 and
column 6 of A(D):
$$(0\ 0\ 0\ 1\ 1\ 0\ 0).(0\ 0\ 0\ 1\ 1\ 0\ 0)^T = (1)(1) + (1)(1) = 2$$
The unit entries in these vectors correspond to the arcs:
     (1,4), (1,5), (4,6) and (5,6) respectively.
The entries are multiplied together as:
$$a_{14}\ a_{46} + a_{15}\ a_{56}$$

Each such product corresponds to a walk of length 2 from 1
to 6, namely the paths:
$$< 1,\ (1,4),\ 4,\ (4,6),\ 6 >$$
$$< 1,\ (1,5),\ 5,\ (5,6),\ 6 >$$
A moment's reflection reveals that scalar multiplications
of this form will yield the number of paths from the new
point (1 in this case) to the column point (6 in this case).
     This is generalized and proven in the following
theorem.

THEOREM 8.1: For any digraph $D = (V, A)$ with adjacency
matrix A(D), each entry a_{ij}^p of $A(D)^p$ equals the number of
walks of length p from point i to point j in D, for any
positive integer p.

PROOF (By induction). A walk of length 1 is a single arc.
Hence each entry a_{ij} in A(D) equals the number of paths of
length 1 from i to j. Thus the theorem is true for p = 1.
Suppose the theorem is true for p = q - 1, (q a positive
integer, $q > 1$).
Then each entry in $A(D)^{q-1}$ equals the number of walks of
length (q-1) between the corresponding points.

     If a_{ij}^{q-1} is the i-j element of $A(D)^{q-1}$ and a_{jk} is the
(j,k) element of A(D) then $a_{ij}^{q-1} a_{jk}$ is the number of
different walks of length (q-1) whose penultimate point is
j, which proceed from point i to point k. Summing over all
points j produces all walks of length q from i to k. But
$$\sum_{j=1}^{n} a_{ij}^{q-1}\ a_{jk}$$
is the (i,k) element in $A(D)^q$. Hence the hypothesis holds
for p = q. This completes the proof by induction.   □
     Besides giving the number of walks of a given length
from one point to another, the matrices $A(D)^q$ can be used
for calculating the distance from one point to another,
since d(i,j) is the smallest q for which a_{ij}^q is non zero,
if such a q exists.

8.3 Digraph Properties and Matrix Properties

In this section we discuss the way in which certain properties of a digraph may be detected in its adjacency matrix.

Firstly, the entries are all 1 or 0. As already remarked, because (v,v) is never an arc, the leading diagonal of A(D) for any digraph D consists of zeros.

Now suppose that the digraph D has

<u>Property 8.1</u> $(v_1,v_2) \in A \Rightarrow (v_2,v_1) \in A$, $v_1,v_2 \in V$ (8.1)

That is, whenever there is an arc from v_1 to v_2, there is an arc from v_2 to v_1. An example of a digraph with this property is shown in Figure 8.2. Digraphs like these are often used to represent communication networks, such as a system of two-way roads. Such digraphs are termed <u>symmetric</u> (see Definition 1.13) and represent a special class of digraphs called simply <u>graphs</u>. Graphs will be discussed in the last chapter of this book. Let us look at the adjacency matrix of the digraph in Figure 8.2.

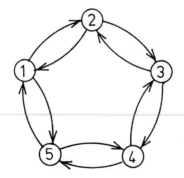

FIGURE 8.2

	1	2	3	4	5
1	0	1	0	0	1
2	1	0	1	0	0
3	0	1	0	1	0
4	0	0	1	0	1
5	1	0	0	1	0

It will be observed that the matrix is also symmetric; that is, symmetrical about its leading diagonal. The condition on a square matrix A to be <u>symmetric</u> is that for each i and j

$$a_{ij} = a_{ji}.$$

Any symmetric matrix of ones and zeros is the adjacency matrix of a digraph in which for given i and j, either both (i,j) and (j,i) are arcs or neither is. Thus a digraph is symmetric precisely if its adjacency matrix is symmetric.

Now suppose that the digraph D has the property that:
$$(v_1, v_2), \in A, \ (v_2, v_3) \in A \text{ and } v_1 \neq v_3 \Rightarrow (v_1, v_3) \in A,$$
$$v_1, v_2, v_3 \in V \quad (8.2)$$

That is, whenever there is an arc from v_1 to v_2 and an arc from v_2 to v_3, then there is an arc from v_1 to v_3. An example of a digraph with this property is shown in Figure 8.3. Such digraphs are termed <u>transitive</u>. (See sections 1.2 and 3.5). Look at the adjacency matrix of the digraph in Figure 8.3:

$$A(D) = \begin{array}{c|ccccc} & 1 & 2 & 3 & 4 & 5 \\ \hline 1 & 0 & 1 & 1 & 1 & 1 \\ 2 & 0 & 0 & 1 & 0 & 0 \\ 3 & 0 & 0 & 0 & 0 & 0 \\ 4 & 0 & 0 & 1 & 0 & 0 \\ 5 & 1 & 1 & 1 & 1 & 0 \end{array}$$

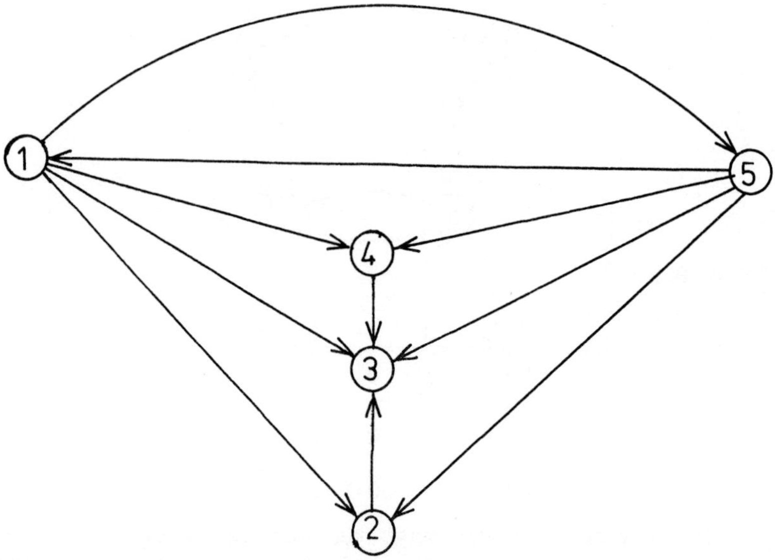

FIGURE 8.3   A transitive digraph

Theorem 8.1 will now be used to identify all paths of length 2 in the digraph. Squaring $A(D)$:

$$A(D)^2 = \begin{bmatrix} 0 & 1 & 1 & 1 & 1 \\ 0 & 0 & 1 & 0 & 0 \\ 0 & 0 & 0 & 0 & 0 \\ 0 & 0 & 1 & 0 & 0 \\ 1 & 1 & 1 & 1 & 0 \end{bmatrix} \begin{bmatrix} 0 & 1 & 1 & 1 & 1 \\ 0 & 0 & 1 & 0 & 0 \\ 0 & 0 & 0 & 0 & 0 \\ 0 & 0 & 1 & 0 & 0 \\ 1 & 1 & 1 & 1 & 0 \end{bmatrix} = \begin{bmatrix} 0 & 1 & 2 & 1 & 0 \\ 0 & 0 & 0 & 0 & 0 \\ 0 & 0 & 0 & 0 & 0 \\ 0 & 0 & 0 & 0 & 0 \\ 0 & 1 & 3 & 1 & 1 \end{bmatrix}$$

Property 8.2:

Whenever there is a positive entry in $A(D)^2$, (except for entries on the leading diagonal) there is a positive entry in the same position in $A(D)$.

Any transitive digraph has this property. This is a direct result of Theorem 8.1 and Property 8.2 since Property 8.2 states that whenever there is a path of length 2 from a point v_1 to a point v_3, there is a path of length 1 from point v_1 to point v_3. Matrices $A(D)^2$ and $A(D)$ display the number of paths of lengths 2 and 1 respectively.

Figure 8.4 shows an acyclic digraph. Its matrix is

	1	2	3	4	5	6
1	0	1	0	0	1	0
2	0	0	1	1	0	0
3	0	0	0	1	0	0
4	0	0	0	0	0	0
5	0	0	0	1	0	0
6	0	0	0	0	1	0

which seems to display nothing in particular except the zero columns 1 and 6, corresponding to the sources and the zero row 4, corresponding to the sink.

But the points have been numbered arbitrarily: if we rearrange the numbers in the order determined by a logical numbering, such as

$$1 \quad 2 \quad 6 \quad 5 \quad 3 \quad 4,$$

then the matrix becomes

	1	2	6	5	3	4
1	0	1	0	1	0	0
2	0	0	0	0	1	1
6	0	0	0	1	0	0
5	0	0	0	0	0	1
3	0	0	0	0	0	1
4	0	0	0	0	0	0

Now all the non-zero entries lie in the part above the leading diagonal.

Property 8.3   A matrix is in upper triangular form if all elements on or below the leading diagonal are zero.

An adjacency matrix in upper triangular form belongs to an acyclic digraph with its points ordered according to a logical numbering. A digraph is acyclic if and only if the rows and columns of its adjacency matrix can be reordered by the same permutation so that the matrix is in upper triangular form.

8.4   The incidence matrix

An arc (v_1, v_2) in a digraph, $D = (V, A)$ is said to be incident away from v_1 and incident towards v_2. Given the digraph D, its incidence matrix I(D), is defined by

$$I(D) = (b_{ij}) \qquad \begin{aligned} i &= 1, 2, \ldots, n, \\ j &= 1, 2, \ldots, m, \end{aligned}$$

where n = the number of points in D,
      m = the number of arcs in D,

$$b_{ij} = \begin{cases} 1 & \text{if arc } j \text{ is incident away from} \quad , \\ -1 & \text{if arc } j \text{ is incident towards} \quad , \\ 0 & \text{otherwise} \end{cases}$$

Notice that this definition of the b_{ij} assumes that the arcs of D are labelled: $1, 2, \ldots, m$. Thus I(D) is an $n \times m$ matrix with as many rows as there are points in D and as many columns as there are arcs in D. For each arc in D there is a "1" and a "-1" the column of the arc, in the positions corresponding to its incident points. The incidence matrix below corresponds to the graph in Figure 8.5(a) where each arc is labelled with a circled number.

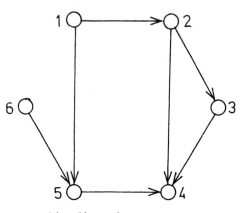

FIGURE 8.4   An acyclic digraph

		1	2	3	4	5	6	7	8
					Arcs				
	1	1	0	0	-1	0	0	0	0
	2	-1	0	0	0	0	0	0	0
	3	0	-1	0	0	-1	1	0	0
points	4	0	1	0	0	0	0	1	0
	5	0	0	-1	0	0	0	0	0
	6	0	0	0	1	0	0	0	0
	7	0	0	0	0	1	0	0	-1
	8	0	0	0	0	0	-1	-1	1
	9	0	0	1	0	0	0	0	0

We can deduce a number of properties of any digraph by examining its incidence matrix. For instance, each column in I(D) has all zeros except for a single "1" entry and a single "-1" entry. This is because each arc is incident away from a single point and is incident toward exactly one other point. These two points are different as any digraph does not contain any loops.

Each row of I(D) corresponds to a point of D. For each arc incident away from (or toward) a point there is a "1" (or a "-1") entry in the row of that point.
So an isolated point will have a corresponding row of all zeros. A pendant point, being incident with just one arc will have exactly one non-zero entry in its corresponding

row. Points: 1,5,6 and 9 illustrate this in Figure 8.5.
     Notice that this digraph is disconnected and has
three weak components. We can partition its set V into
subsets of connected points: {1,2,6} {3,4,7,8}, {5,9}. We
can then partition its set A into corresponding subsets:
{1,4}, {2,5,6,7,8},{3}.

Suppose we now rewrite the matrix, ordering the rows and
columns as they are shown in the above partitions:

	1	2	3	4	5	6	7	8
1	1	-1						
2	-1	0						
3	0	1						
4			-1	-1	1	0	0	
5			1	0	0	1	0	
6			0	1	0	0	-1	
7			0	0	-1	-1	1	
8								-1
9								1

The relabelled digraph is shown in Figure 8.5(b).
     The blank entries are all zeros. It can be seen that
this re-ordered matrix is <u>block-diagonal</u>, i.e. it can be
expressed as a set of non-zero diagonal submatrices, all
other submatrices being zero. Each non-zero diagonal
submatrix corresponds to one of the weak components of the
digraph.

8.5 Exercises

S 1. Given the following adjacency matrices, draw arrow
     diagrams of the corresponding digraphs. Which are
     symmetric? Which are transitive? Which are acyclic?

$$
\begin{bmatrix}
0 & 0 & 0 & 0 & 0 \\
0 & 0 & 0 & 0 & 0 \\
0 & 0 & 0 & 0 & 0 \\
0 & 0 & 0 & 0 & 0 \\
0 & 0 & 0 & 0 & 0
\end{bmatrix}
\qquad
\begin{bmatrix}
0 & 0 & 0 & 0 & 0 \\
1 & 0 & 0 & 0 & 0 \\
1 & 1 & 0 & 0 & 0 \\
1 & 1 & 1 & 0 & 0 \\
1 & 1 & 1 & 1 & 0
\end{bmatrix}
$$

          (i)   (ii)

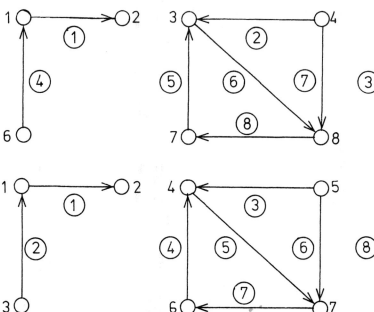

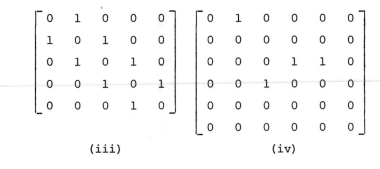

FIGURE 8.5 The relabelling of a digraph

$$\begin{bmatrix} 0 & 1 & 0 & 0 & 0 \\ 1 & 0 & 1 & 0 & 0 \\ 0 & 1 & 0 & 1 & 0 \\ 0 & 0 & 1 & 0 & 1 \\ 0 & 0 & 0 & 1 & 0 \end{bmatrix} \quad \begin{bmatrix} 0 & 1 & 0 & 0 & 0 & 0 \\ 0 & 0 & 0 & 0 & 0 & 0 \\ 0 & 0 & 0 & 1 & 1 & 0 \\ 0 & 0 & 1 & 0 & 0 & 0 \\ 0 & 0 & 0 & 0 & 0 & 0 \\ 0 & 0 & 0 & 0 & 0 & 0 \end{bmatrix}$$

(iii)   (iv)

S 2. Construct the diagram of the digraph with the following adjacency matrix and calculate the number of walks of length 3 from point 2 to point 4. Verify that this number is correct by finding all paths from your diagram.

$$\begin{bmatrix} 0 & 0 & 1 & 1 \\ 1 & 0 & 1 & 0 \\ 1 & 0 & 0 & 1 \\ 0 & 0 & 0 & 0 \end{bmatrix}$$

S 3. Construct the adjacency matrices of the digraphs shown
     in Figure 8.6.

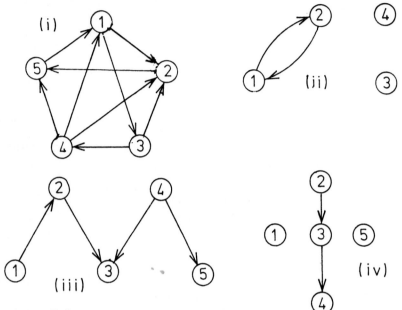

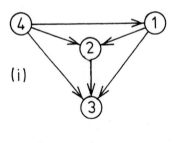

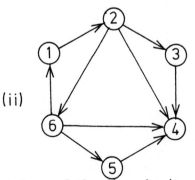

FIGURE 8.6

S 4. Verify whether either of the digraphs in Figure 8.7
     is transitive by comparing $A(D)$ and $[A(D)]^2$ in each
     case.

FIGURE 8.7

S 5. Construct the incidence matrices of the digraphs in
     Figure 8.8

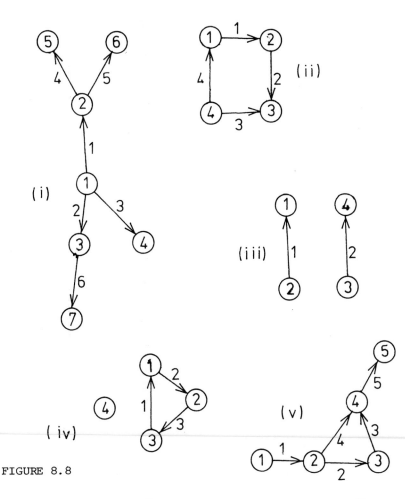

FIGURE 8.8

6. What can be said about the nth power of the adjacency matrix of an acyclic digraph, if n is large enough? How large is 'large enough'?

S 7. If T is a rooted tree what is the set of values which may appear as entries in $A(T)^n$ for any n?

* 8. Find a digraph D such that $A(D)^3 = A(D)$. Repeat for $A(D)^4 = A(D)$, $A(D)^5 = A(D)$ etc. Can there be a digraph such that $A(D)^2 = A(D)$?

S 9. Given the following incidence matrices, draw diagrams of the corresponding digraphs:

G

$$\begin{bmatrix} 1 & -1 & 0 & 0 \\ 0 & 1 & -1 & 0 \\ -1 & 0 & 1 & -1 \\ 0 & 0 & 0 & 1 \end{bmatrix} \qquad \begin{bmatrix} -1 & -1 & -1 & -1 \\ 1 & 0 & 0 & 0 \\ 0 & 1 & 0 & 0 \\ 0 & 0 & 1 & 0 \\ 0 & 0 & 0 & 1 \end{bmatrix}$$

$$(i) \qquad\qquad\qquad (ii)$$

8.6   The cycle matrix

Recall that a cycle in a digraph $D = (V,A)$ is a closed
walk in which all points are distinct except for the first
and last (which must be equal for the walk to be closed).
(Definition 1.7). So for instance, the cycles in the
digraph in Figure 8.9(a) are:
$< 1, (1,2), 2, (2,3), 3, (3,4), 4, (4,5), 5, (5,1), 1 >$,
$< 1, (1,4), 4, (4,5), 5, (5,1), 1 >$
$< 2, (2,3), 3, (3,4), 4, (4,2), 2 >$

The arcs in the figure have been labelled with the
circled numbers as shown. Let us now remove the arrows from
the arcs of this digraph. We call this process the
disorientation of the digraph: the result is shown in
Figure 8.9(b). The structure produced is called a
multigraph. If (u,v) and (v,u) are both arcs then u and v
are linked by two lines. When there are no such pairs of
arcs, as in the example treated here, the disoriented
structure is a graph, as discussed in Chapter 9.
There are a number of cycles in the graph in Figure 8.9(b):

1: $<1, \{1,2\}, 2, \{2,4\}, 4, \{4,1\}, 1 >$,
2: $<1, \{1,4\}, 4, \{4,5\}, 5, \{5,1\}, 1 >$,
3: $<2, \{2,4\}, 4, \{4,3\}, 3, \{3,2\}, 2 >$,
4: $<1, \{1,2\}, 2, \{2,3\}, 3, \{3,4\}, 4, \{4,1\}, 1 >$,
5: $<1, \{1,2\}, 2, \{2,4\}, 4, \{4,5\}, 5, \{5,1\}, 1 >$,
6: $<1, \{1,2\}, 2, \{2,3\}, 3, \{3,4\}, 4, \{4,5\}, 5, \{5,1\}, 1 >$

(In a multigraph cycle a line may be traversed in either
direction, but may not be used more than once).
We can give each cycle either a clockwise or anti-clockwise
orientation. The orientation we assign to each cycle is
arbitrary, so it may be defined by the definitions of the
cycles above. These directions are indicated in
Figure 8.9(c).
Now each arc in the original digraph in Figure 8.9(a)
either has the same orientation as a cycle of which it is

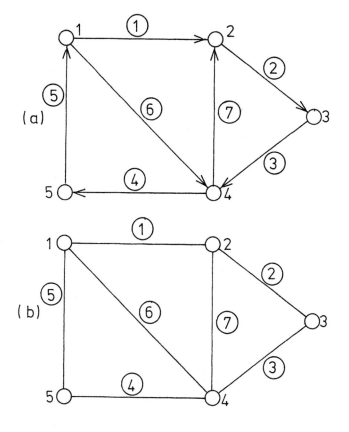

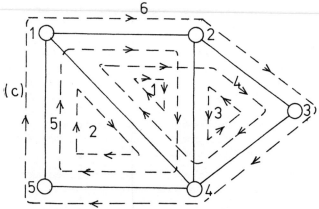

FIGURE 8.9   The disorientation of a digraph

part, or the opposite orientation e.g. arc (2,3) has the
same orientation as cycle 4, but the opposite orientation
to cycle 3. We can express all the information of this
form in the matrix.

arcs

		1	2	3	4	5	6	7
	1	1	0	0	0	0	-1	-1
	2	0	0	0	1	1	1	0
	3	0	-1	-1	0	0	0	-1
cycles	4	1	1	1	0	0	-1	0
	5	1	0	0	1	1	0	-1
	6	1	1	1	1	1	0	0

Each column corresponds to an arc, each row to a cycle. We
can build up the matrix row by row. In each row a 1 is
entered in each position where the associated arc belongs
to the cycle of the row and has the same orientation, e.g.
arc 1 belongs to cycle 1 and has the same orientation; so
the (1,1) entry is "1". A "-1" is entered in each position
where the associated arc belongs to the cycle of the row
and has the opposite orientation, e.g. arc 6 belongs to
cycle 1 and has the opposite orientation; so that the
(1,6) entry is "-1". Similarly for the (1,7) entry. A
zero is entered in each position where the associated arc
does not belong to the cycle of the row, e.g. arcs 2, 3, 4
and 5 do not belong to cycle 1, so zeros are entered in
entries: (1,2), (1,3), (1,4) and (1,5). This process has
filled up row 1 in the above matrix. The other rows are
filled in the same manner.
    Let us now generalize the definition of matrices of
this type. Given a digraph $D = (V,A)$, its cycle matrix.
$C(D)$, is defined by:

$$C(D) = (c_{ij}) \quad , \quad \begin{array}{l} i = 1, 2, \ldots, p, \\ j = 1, 2, \ldots, m, \end{array}$$

where p = the number of cycles present in the
                    multigraph produced by disorientating D,
                m = the number of arcs of D,

$$c_{ij} = \begin{cases} 1, & \text{if the } i^{th} \text{cycle includes arc } j \text{ with} \\ & \text{the same orientation,} \\ -1, & \text{if the } i^{th} \text{ cycle includes arc } j \text{ with} \\ & \text{opposite orientation,} \\ 0 & \text{otherwise.} \end{cases}$$

Thus C(D) is a p × m matrix with as many rows as there are
cycles in the disoriented graph of D, and as many columns
as there are arcs in D. As was previously stated, the
orientation of each cycle is arbitrary. If the
orientation of any cycle is reversed, this simply changes
the sign of each non-zero element in the row
corresponding to this cycle in C(D).

We can deduce a number of properties of any digraph by
examining its cycle matrix. For instance, the number of
non-zero entries in any column indicates the number of
cycles to which the corresponding arc belongs. Hence an
arc with a column of all zeros does not belong to any
cycle. The number of non-zero entries in any row
indicates the number of arcs which make up the cycle
corresponding to that row.

As in the case of the incidence matrix, the cycle
matrix can be partitioned into non-zero diagonal
submatrices which correspond to connected components of the
digraph. For this reason we assume that any digraph for
which the cycle matrix is constructed is connected.

If any two rows or any two columns in C(D) are
interchanged this corresponds to a relabelling of the rows,
or columns involved.

Of course a digraph which is disoriented may produce
no cycles. Hence its cycle matrix does not exist. This
means that unlike the adjacency or incidence matrices, the
cycle matrix does not completely define a digraph. Rooted
trees, for example, have no cycle matrix.

The incidence matrix of the digraph in Figure 8.9(a)
is:

$$
I(D) = \begin{array}{c|ccccccc}
 & 1 & 2 & 3 & 4 & 5 & 6 & 7 \\
\hline
1 & 1 & 0 & 0 & 0 & -1 & 1 & 0 \\
2 & -1 & 1 & 0 & 0 & 0 & 0 & -1 \\
3 & 0 & -1 & 1 & 0 & 0 & 0 & 0 \\
4 & 0 & 0 & -1 & 1 & 0 & -1 & 0 \\
5 & 0 & 0 & 0 & -1 & 1 & 0 & 0 \\
\end{array}
$$

We shall now examine examples of five different types of
relationships between a cycle and a point.

(i)   The point is not part of the cycle.
      Consider cycle 1 and point 3. Point 3 is not part of
      cycle 1. Let us consider the row from C(D)
      corresponding to cycle 1 and the row from I(D)
      corresponding to point 1:

$(1 \ 0 \ 0 \ 0 \ 0 \ -1 \ -1) \cdot (0 \ -1 \ 1 \ 0 \ 0 \ 0 \ 0)^T = 0$.
When we take the transpose (denoted by "T") of the
vector corresponding to the row from $I(D)$ and multiply
the two together the resulting scalar obtained is zero.
This should come as no surprise as the arcs of the
cycle and the arcs incident with point 3 form two
disjoint sets. This means that the vectors never both
have a non-zero element in the same position.
        If a point does occur in a cycle it is incident
with two sequential arcs of the cycle. There are then
four cases to investigate.

(ii) Two sequential arcs of a cycle, only the second having
      the cycle orientation.
      Consider cycle 1 and point 1. The arcs of cycle 1
      which are incident with point 1 are: (4,1) and (1,2).
      Arc (4,1) has the opposite orientation to cycle 1,
      while (1,2) has the same orientation. On multiplying
      the cycle 1, point 1 vectors together as before;
      $(1 \ 0 \ 0 \ 0 \ 0 \ -1 \ -1) \cdot (1 \ 0 \ 0 \ 0 \ -1 \ 1 \ 0)^T = 0$
      We see that their product is once again zero.

(iii) Two sequential arcs of a cycle, only the first having
      the cycle orientation.
      Consider cycle 4 and point 4. The arcs of cycle 4
      which are incident with point 4 are: (3,4) and (4,1).
      Arc (3,4) has the same orientation as cycle 4, while
      (4,1) has the opposite orientation. On multiplying
      the cycle 4, point 4 vectors together as before;
      $(1 \ 1 \ 1 \ 0 \ 0 \ -1 \ 0) \cdot (0 \ 0 \ -1 \ 1 \ 0 \ -1 \ 1)^T = 0$
      We see that their product is once again zero.

(iv) Two sequential arcs of a cycle, both having the cycle
      orientation.
      Consider cycle 2 and point 5. The arcs of cycle 2
      which are incident with point 5 are: (4,5) and (5,1).
      Both have the same orientation as cycle 2. On
      multiplying the cycle 2, point 5 vectors together as
      before;
      $(0 \ 0 \ 0 \ 1 \ 1 \ 1 \ 0) \cdot (0 \ 0 \ 0 \ -1 \ 1 \ 0 \ 0)^T = 0$
      We see that their product is once again zero.

(v)   Two sequential arcs of a cycle, neither having the
      cycle orientation.
      Consider cycle 3 and point 3. The arcs of the cycle
      which are incident with point 4 are: (2,3) and (3,4).
      Both have the opposite orientation to cycle 3. On
      multiplying the cycle 3, point 3 vectors together:
      $(0 \ -1 \ -1 \ 0 \ 0 \ 0 \ -1) \cdot (0 \ -1 \ 1 \ 0 \ 0 \ 0 \ 0)^T = 0$.
      Hence in all five cases multiplying a row of $C(D)$ with
a transposed row of $I(D)$ results in a product of zero. It

turns out that if we multiply any such pair of rows from
C(D) and I(D) for any digraph, D the product is zero. This
can be summarized in the form of a theorem which we state
without proof:

THEOREM 8.2:   Given any digraph, D for which there exists a
cycle matrix, C(D), then if I(D) is the incidence matrix
of D:

$$C(D)I(D)^T = 0$$

8.7   The fundamental cycle matrix
Recall the cycle matrix, C(D) presented for the
digraph in Figure 8.9(a):

	1	2	3	4	5	6	7
1	1	0	0	0	0	-1	-1
2	0	0	0	1	1	1	0
3	0	-1	-1	0	0	0	-1
4	1	1	1	0	0	-1	0
5	1	0	0	1	1	0	-1
6	1	1	1	1	1	0	0

If rows 1 and 2 are added together, row 5 is produced.
Thus the rows of C(D) are linearly dependent. This
operation of adding rows together corresponds graphically
to combining cycles to generate a new cycle. This is
illustrated in Figure 8.10 with cycles: 1, 2 and 5. In
this figure, the cycles are represented and not the arcs.
Note that cycles 1 and 2 both contain an arc linking
points 1 and 6. As these two arcs are orientated in
opposite directions, they cancel out when the two cycles
are added together; cycle 5 does not contain an arc
between points 1 and 6. This cancelling out process is
reflected in the row addition operation, for example:

$$c_{16} + c_{26} = 0 = c_{56}$$

Hence it has been shown that we can build cycle 5 from
cycles 1 and 2. Can we find a minimum number of
fundamental cycles which can be used to build all other
cycles? [This is especially handy for large digraphs
which are likely to have many cycles]. We assume that any
number of fundamental cycles can be added or subtracted to
produce one of the other cycles. For instance let us add
rows 1 and 2 and subtract row 3:

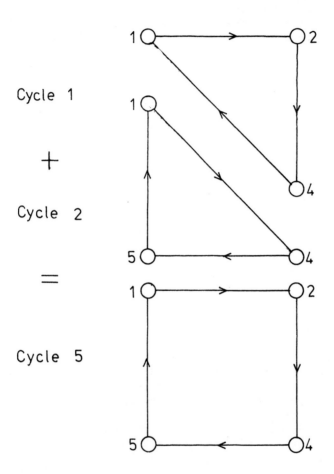

FIGURE 8.10   The addition of cycles

$$
\begin{array}{llll}
\text{row 1:} & & (1 & 0 & 0 & 0 & 0 & -1 & -1) \\
\text{row 2:} & + & (0 & 0 & 0 & 1 & 1 & 1 & 0) \\
\text{row 3:} & - & \underline{(0 & -1 & -1 & 0 & 0 & 0 & -1)} \\
& & (1 & 1 & 1 & 1 & 1 & 0 & 0)
\end{array}
$$

This new vector corresponds to cycle 6.   (The reader should verify this result graphically, using the method pictured in Figure 8.10).
If we are going to extract a minimal set of fundamental cycles it will clearly be impracticable to try all

combinations. We need an algorithm.

We now outline the process by which a minimal set of fundamental cycles can be identified.

A spanning tree $T = (V_1, E_1)$ of a multigraph (V_2, E_2) is a graph in which $V_1 = V_2$ and $E_1 \subseteq E_2$, and there exists a unique path between every pair of points in V. (See Chapter 9, particularly Section 9.6 for details of graphtheoretic definitions). As an example the graph in Figure 8.11(a) is a spanning tree for the graph in Figure 8.9(b). This is so because this subgraph contains all the points of the original graph, it is possible to find a path from every point to every other point, and there are no cycles. Having found a spanning tree we now identify all the lines of the original graph which are not members of the spanning tree. These lines are called chords and are shown as heavy lines in Figure 8.11(b), the lines of the spanning tree being shown as dashed lines. The lines of a spanning tree are termed branches.

It can be seen that each chord defines a cycle which consists of the chord and lines from the spanning tree. For instance line 2 defines a cycle with lines 3 and 7, line 2 being the only chord. And line 4 defines a cycle with lines 5 and 6, line 4 being the only chord. Next we assign an orientation to each of these cycles which is the same as that of its chord. (Recall that each chord was originally an arc). So on finding the orientation of the chords from Figure 8.9(a) we can assign the cycle orientations as shown in Figure 8.11(c).

The cycles found in this way are a set of fundamental cycles. We can build any of the other cycles shown in Figure 8.9(c) from them. Let us label these three cycles: α, β and γ as shown in Figure 8.11(c). Then the fundamental cycle matrix is:

	1	2	3	4	5	6	7
α	1	0	0	0	0	-1	-1
β	0	0	0	1	1	1	0
γ	0	1	1	0	0	0	1

Then on comparing this matrix with the cycle matrix for the graph in Figure 8.9(a) given in the last section, it can be seen that:

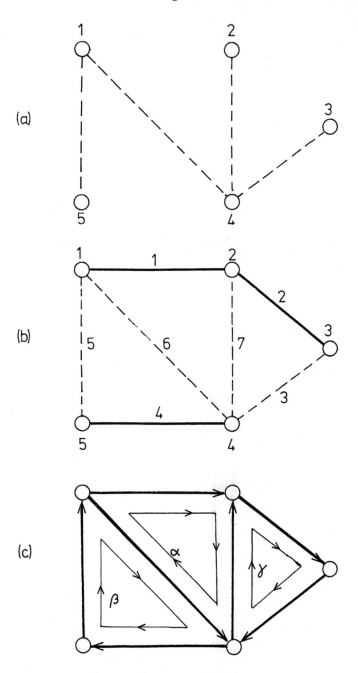

Figure 8.11 Identifying fundamental circuits

$$\alpha \; = \; (1 \quad 0 \quad 0 \quad 0 \quad 0 \quad -1 \quad -1) \; = \; \text{cycle 1}$$

$$\beta \; = \; (0 \quad 0 \quad 0 \quad 1 \quad 1 \quad 1 \quad 0) \; = \; \text{cycle 2}$$

$$-\;\gamma \; = \; (0 \quad -1 \quad -1 \quad 0 \quad 0 \quad 0 \quad -1) \; = \; \text{cycle 3}$$

$$
\begin{aligned}
\alpha + \gamma \; = \; &(1 \quad 0 \quad 0 \quad 0 \quad 0 \quad -1 \quad -1) \\
+ \; &(0 \quad 1 \quad 1 \quad 0 \quad 0 \quad 0 \quad 1) \\
\hline
= \; &(1 \quad 1 \quad 1 \quad 0 \quad 0 \quad -1 \quad 0) \; = \; \text{cycle 4}
\end{aligned}
$$

$$
\begin{aligned}
\alpha + \beta \; = \; &(1 \quad 0 \quad 0 \quad 0 \quad 0 \quad -1 \quad -1) \\
+ \; &(0 \quad 0 \quad 0 \quad 1 \quad 1 \quad 1 \quad 0) \\
\hline
&(1 \quad 0 \quad 0 \quad 1 \quad 1 \quad 0 \quad -1) \; = \; \text{cycle 5}
\end{aligned}
$$

$$
\begin{aligned}
\alpha + \beta + \gamma \; = \; &(1 \quad 0 \quad 0 \quad 0 \quad 0 \quad -1 \quad -1) \\
+ \; &(0 \quad 0 \quad 0 \quad 1 \quad 1 \quad 1 \quad 0) \\
+ \; &(0 \quad 1 \quad 1 \quad 0 \quad 0 \quad 0 \quad 1) \\
\hline
&(1 \quad 1 \quad 1 \quad 1 \quad 1 \quad 0 \quad 0) \; = \; \text{cycle 6}
\end{aligned}
$$

The reader should verify these constructions graphically, using Figure 8.9(c).   It can be shown by induction that if a graph has n points then any of its spanning trees will have n-1 lines.   Thus if the graph has m lines altogether it will have

$$m - (n-1) = m - n + 1 \quad \text{chords.}$$

As each chord defines a fundamental cycle, there will be $(m - n + 1)$ such fundamental cycles.   If they are labelled $\alpha_1, \alpha_2, \ldots, \alpha_{m-n+1}$ we can formally define the <u>fundamental cycle matrix</u>, $F(D)$ of a digraph D as:

$$F(D) = (f_{ij}) \qquad \begin{aligned} i &= 1,2,\ldots,m - n + 1, \\ j &= 1,2,\ldots,m \end{aligned}$$

where

$$
f_{ij} = \begin{cases}
1 \text{ , if the } i^{th} \text{ fundamental cycle includes arc j} \\
\quad \text{with the same orientation,} \\
-1 \text{ , if the } i^{th} \text{ fundamental cycle includes arc j} \\
\quad \text{with the opposite orientation,} \\
0 \text{ , otherwise.}
\end{cases}
$$

Of course not all linear combinations of fundamental cycles produce new cycles.   For instance

$$
\begin{aligned}
\beta + \gamma \; = \; &(0 \quad 1 \quad 0 \quad 0 \quad 1 \quad 1 \quad 0) \\
+ \; &(0 \quad 0 \quad 1 \quad 1 \quad 0 \quad 0 \quad 1) \\
\hline
&(0 \quad 1 \quad 1 \quad 1 \quad 1 \quad 1 \quad 1)
\end{aligned}
$$

This new vector does not represent a cycle. However, the
point is that all cycles can be constructed by forming
linear combinations of the fundamental cycles.

In the example just discussed the spanning tree chosen
was not the only one available. In most digraphs there is
usually a large number of spanning trees to choose from.
Each different spanning tree creates a different set of
chords. Each set of chords creates a different set of
fundamental cycles. However, any of these sets of
fundamental cycles is capable of building all the cycles of
the digraph.

Let us re-order the columns in the fundamental cycle
matrix of the example, by placing the chords 1,4 and 2
first:

	1	4	2	5	6	7	3
α	1	0	0	0	-1	-1	0
β	0	1	0	1	1	0	0
γ	0	0	1	0	0	1	1

It can be seen that this matrix can be split into two parts,
one of them being the identity matrix:

$$\begin{bmatrix} 1 & 0 & 0 & & 0 & -1 & -1 & 0 \\ 0 & 1 & 0 & & 1 & 1 & 0 & 0 \\ 0 & 0 & 1 & & 0 & 0 & 1 & 1 \end{bmatrix}$$

The fact that the left hand matrix is the identity matrix
should be no surprise. Recall that we carefully
constructed each fundamental cycle to contain a unique
chord. Hence F(D) can be partitioned into:

$$F(D) = [\,I : \bar{F}\,(D)\,]$$

where I is the identity matrix. In the above example:

$$\bar{F}\,(D) \quad = \quad \begin{bmatrix} 0 & -1 & -1 & 0 \\ 1 & 1 & 0 & 0 \\ 0 & 0 & 1 & 1 \end{bmatrix}$$

The matrix $\bar{F}$ (D) will be discussed in the next section

8.8   The fundamental cut-set matrix

Consider the digraph in Figure 8.9(a). If arc 1 is
removed the digraph is still weakly connected. In other
words there is still a walk between every pair of vertices
when the digraph is disoriented. Now suppose arc 1 is

replaced and arcs 2 and 3 are removed. In this case the
digraph has been split into two disconnected pieces:
point 3 in one piece and the rest in the other. In other
words there are no walks to or from point 3 when the
digraph is disoriented. Now suppose arcs 2 and 3 are
replaced and arcs 1, 4, 5 and 6 are removed. In this case
the digraph has been split into three disconnected pieces:
point 1; point 5; and points 2, 3 and 4 together with arcs
2, 3 and 7. We are interested in finding sets of arcs whose
removal splits the digraph into exactly two pieces. The
additional proviso is stipulated that each set of arcs must
be minimal in the sense that the removal of no proper
subset of the set also splits the digraph into two sets.
Such minimal disconnecting sets are called cut-sets.
(Some books use the term proper cut set).

As an illustration of this consider the removal of
arcs 1, 3, 5 and 6 from the digraph in Figure 8.9(a). This
removal splits the graph into two pieces: point 1 and the
rest. This set is not minimal as arc 3 is not needed to
disconnect the digraph. Hence this set is not a cut set.
However the set of arcs 1, 5 and 6 is a cut set. In fact
the complete list of cut sets of the digraph is:

cut-set	arcs
1	1,5,6
2	4,5
3	1,2,7
4	3,4,6,7
5	2,3
6	1,4,6
7	2,5,6,7
8	3,5,6,7
9	2,4,6,7
10	1,3,7

We shall now attempt to find a set of fundamental
cut-sets, from which all cut-sets can be constructed. This
whole idea is very similar to that of the set of
fundamental cycles.

Once again we begin by identifying a spanning tree of
the disoriented graph of the digraph. For our example, the
tree found in Figure 8.11(a) will do. On examining this
tree (without any of the chords present) we see that if any

one of its lines is removed the subgraph remaining is
broken into two parts. That is on looking at Figure
8.11(a):

The removal of line:		creates the connected graphs with	
5	points 5	and	1,2,3,4
6	1,5		2,3,4
7	2		1,3,4,5
3	3		1,2,4,5

As has just been stated, the removal of line 5 from the
tree in Figure 8.11(a) splits the tree into subtrees:
one tree is just point 5 with no lines; the other is
points; 1,2,3 and 4 with lines 6, 7 and 3. What chords
connect these two subtrees? It can be seen from Figure
8.5(b) that the chord 4 links points 4 and 5. The set of
arcs which corresponds to this branch and its connecting
chords (just 4 in this case) is $\{5,4\}$. This set is a cut
set for the digraph in Figure 8.9(a). We can create such a
set for each of the branches in Figure 8.11(a).

$$5 \quad \sim \quad \{ \ 5, \quad 4 \quad \},$$
$$6 \quad \sim \quad \{ \ 6, \quad 1, \quad 4 \quad \},$$
$$7 \quad \sim \quad \{ \ 7, \quad 1, \quad 2 \quad \},$$
$$3 \quad \sim \quad \{ \ 3, \quad 2 \quad \}$$

These cut-sets make up a set of fundamental cut-sets for
the digraph. Let us reconsider the first of these cut
sets: $\{5, 4\}$. Remember this cut-set breaks the digraph up
into two digraphs:

digraph (a)   :   point 5
digraph (b)   :   points 1, 2, 3 and 4 together with arcs
                  1, 2, 3, 6 and 7.

Now the branch which created this cut-set, namely arc 5, is
orientated from digraph (a) to digraph (b)   (see Figure
8.9(a)). However, the only chord in the cut-set (arc 4) is
orientated the opposite way, from digraph (b) to digraph
(a). We can summarize this information as follows:

	1	2	3	4	5	6	7
α	0	0	0	-1	1	0	0

The α indicates that the information is about cut-set α.
The numbers 1, 2,..., 7 represent the arcs. The 1 entry in
column 5 indicates the unique branch of the cut-set. If
there were other arcs with the same orientation (from
digraph (a) to digraph (b)), they would also have a 1
entered. A (-1) entry indicates arcs in the cut-set with
the opposite orientation (from digraph (b) to digraph (a))

from the branch.   Zero entries represent arcs which are not part of the cut-set.

The rows for the other fundamental cut-sets can be filled in by the same process:

	1	2	3	4	5	6	7
α	0	0	0	-1	1	0	0
β	1	0	0	-1	0	1	0
γ	1	-1	0	0	0	0	1
δ	0	-1	1	0	0	0	0

Let us now generalize the definition of matrices of this type.   Given a digraph $D = (V,A)$, its <u>fundamental cut-set matrix</u> $K(D)$ is defined by:

$$K(D) = (k_{ij}) \qquad \begin{array}{l} i = 1,2,\ldots,(n-1), \\ j = 1,2,\ldots,m \end{array}$$

$$\text{where } n = \text{the number of points of } D,$$
$$m = \text{the number of arcs of } D,$$

$$k_{ij} \begin{cases} 1 \text{ , if the } i^{th} \text{ fundamental cut-set includes the } j^{th} \\ \quad \text{arc with the same orientation as the branch of} \\ \quad \text{this set,} \\ -1 \text{ , of the } i^{th} \text{ fundamental cut-set includes the } j^{th} \\ \quad \text{arc with the opposite orientation to the branch} \\ \quad \text{of this set,} \\ 0 \text{ , otherwise} \end{cases}$$

As has been mentioned before any spanning tree of a digraph with n points contains (n-1) lines.   Each branch of a spanning tree defines exactly one fundamental cut-set. Hence there are (n-1) fundamental cut-sets.

We can deduce a number of properties of any digraph, D by examining $K(D)$.   For instance, the number of non-zero entries in any column indicates the number of fundamental cut-sets to which the corresponding arc belongs.   Hence an arc with a column of all zeros does not belong to any fundamental cut-set.   The number of non-zero entries in any row indicates the number of arcs which make up the cut-set corresponding to that row.   If any two rows or any two columns of $K(D)$ are interchanged, this corresponds to a relabelling of the rows or columns involved.   Consider the following subtraction:

$$\alpha - \beta = \quad (0 \quad 0 \quad 0 \quad -1 \quad 1 \quad 0 \quad 0)$$

$$-(1 \quad 0 \quad 0 \quad -1 \quad 0 \quad 1 \quad 0)$$

$$= \quad (-1 \quad 0 \quad 0 \quad 0 \quad 1 \quad -1 \quad 0)$$

This subtract produces a cut-set of the arcs: 1,5 and 6.
Our convention concerning the positive and negative entries
still holds. This cut-set splits the digraph into two
parts: (a)   point 1 only in one part and (b) points 2, 3,
4 and 5 together with arcs 2, 3, 4 and 7 in the other.
(See Figure 8.5(c)). The branch in this cut-set with a
positive entry is arc 5.   It is orientated from the first
digraph (a) to digraph (b). The other arcs are orientated
in the opposite direction and hence have a (-1) entry. We
could of course have made the following subtraction:

$$\beta - \alpha = \quad (1 \quad 0 \quad 0 \quad -1 \quad 0 \quad 1 \quad 0)$$

$$-(0 \quad 0 \quad 0 \quad -1 \quad 1 \quad 0 \quad 0)$$

$$(1 \quad 0 \quad 0 \quad 0 \quad -1 \quad 1 \quad 0)$$

In this case arc 6 is the branch with a positive entry and
is orientated from digraph (b) to digraph (a). Hence the
signs are reversed in the vector corresponding to this cut-
set.

   We shall now construct all the cut-sets of the table
3.1 from the fundamental cut-sets:

$$\alpha - \beta = (0 \quad 0 \quad 0 \quad -1 \quad 1 \quad 0 \quad 0)$$

$$-(1 \quad 0 \quad 0 \quad -1 \quad 0 \quad 1 \quad 0)$$

$$= (-1 \quad 0 \quad 0 \quad 0 \quad 1 \quad -1 \quad 0) = \text{cut-set 1,}$$

$$\alpha = (0 \quad 0 \quad 0 \quad -1 \quad 1 \quad 0 \quad 0) = \text{cut-set 2,}$$

$$\gamma = (1 \quad -1 \quad 0 \quad 0 \quad 0 \quad 0 \quad 1) = \text{cut-set 3,}$$

$$\beta - \gamma + \delta = (1 \quad 0 \quad 0 \quad -1 \quad 0 \quad 1 \quad 0)$$

$$-(1 \quad -1 \quad 0 \quad 0 \quad 0 \quad 0 \quad 1)$$

$$+(0 \quad -1 \quad 1 \quad 0 \quad 0 \quad 0 \quad 0)$$

$$(0 \quad 0 \quad 1 \quad -1 \quad 0 \quad 1 \quad -1) = \text{cut-set 4,}$$

$$\delta = (0 \quad -1 \quad 1 \quad 0 \quad 0 \quad 0 \quad 0) = \text{cut-set 5,}$$

$$\beta = (1 \quad 0 \quad 0 \quad -1 \quad 0 \quad 1 \quad 0) = \text{cut-set 6,}$$

$$-\alpha + \beta - \gamma = -(0 \quad 0 \quad 0 \quad -1 \quad 1 \quad 0 \quad 0)$$

$$ +(1 \quad 0 \quad 0 \quad -1 \quad 0 \quad 1 \quad 0)$$

$$ \underline{-(1 \quad -1 \quad 0 \quad 0 \quad 0 \quad 0 \quad 1)}$$

$$ (0 \quad 1 \quad 0 \quad 0 \quad -1 \quad 1 \quad -1) = \text{cut-set 7,}$$

$$\alpha - \beta + \gamma - \delta = (0 \quad 0 \quad 0 \quad -1 \quad 1 \quad 0 \quad 0)$$

$$ -(1 \quad 0 \quad 0 \quad -1 \quad 0 \quad 1 \quad 0)$$

$$ +(1 \quad -1 \quad 0 \quad 0 \quad 0 \quad 0 \quad 1)$$

$$ \underline{-(0 \quad -1 \quad 1 \quad 0 \quad 0 \quad 0 \quad 0)}$$

$$ (0 \quad 0 \quad -1 \quad 0 \quad 1 \quad -1 \quad 1) = \text{cut-set 8,}$$

(Note in this case there are two branches (arcs 5 and 7) which specify the cut-set orientation).

$$\beta - \gamma = (1 \quad 0 \quad 0 \quad -1 \quad 0 \quad 1 \quad 0)$$

$$ \underline{-(1 \quad -1 \quad 0 \quad 0 \quad 0 \quad 0 \quad 1)}$$

$$ = (0 \quad 1 \quad 0 \quad -1 \quad 0 \quad 1 \quad -1) = \text{cut-set 9.}$$

$$\gamma - \delta = (1 \quad -1 \quad 0 \quad 0 \quad 0 \quad 0 \quad 1)$$

$$ \underline{-(0 \quad -1 \quad 1 \quad 0 \quad 0 \quad 0 \quad 0)}$$

$$ = (1 \quad 0 \quad -1 \quad 0 \quad 0 \quad 0 \quad 1) = \text{cut-set 10.}$$

As with fundamental cycles, not all linear combinations of fundamental cut-sets produce new cut-sets. For instance

$$\alpha + \gamma - \delta = (0 \quad 0 \quad 0 \quad -1 \quad 1 \quad 0 \quad 0)$$

$$ +(1 \quad -1 \quad 0 \quad 0 \quad 0 \quad 0 \quad 1)$$

$$ \underline{-(0 \quad -1 \quad 1 \quad 0 \quad 0 \quad 0 \quad 0)}$$

$$ (1 \quad 0 \quad 1 \quad -1 \quad 1 \quad 0 \quad 1)$$

This vector does not represent a cut-set as the removal of its arcs: 1, 3, 4, 5 and 7 splits the digraph into more than two pieces.

Also, as with fundamental cycles, the spanning tree chosen is arbitrary. Different spanning trees will create different sets of fundamental cut-sets. However any of the sets of fundamental cut-sets produced in this way is capable of building all the cut-sets of the digraph.

Let us reorder the columns of K(D) of the example into the same sequence as that for F(D):

	1	4	2	5	6	7	3
α	0	-1	0	1	0	0	0
β	1	-1	0	0	1	0	0
γ	1	0	-1	0	0	1	0
δ	0	0	-1	0	0	0	1

It can be seen that this matrix, like F(D), can be split into two parts; one of them being the identity matrix:

$$\begin{bmatrix} 0 & -1 & 0 & 1 & 0 & 0 & 0 \\ 1 & -1 & 0 & 0 & 1 & 0 & 0 \\ 1 & 0 & -1 & 0 & 0 & 1 & 0 \\ 0 & 0 & -1 & 0 & 0 & 0 & 1 \end{bmatrix}$$

Once again the fact that the right-hand matrix is the identity matrix should be no surprise: Recall that we carefully constructed each fundamental cut-set to contain exactly one branch of the spanning tree;
Hence K(D) can be partitioned into:

$$K(D) = [\,\overline{K}(D) : I\,].$$

In the above example:

$$K(D) = \begin{bmatrix} 0 & -1 & 0 \\ 1 & -1 & 0 \\ 1 & 0 & -1 \\ 0 & 0 & -1 \end{bmatrix}$$

Now the transpose of $\overline{K}(D)$ is:

$$\overline{K}(D)^T = \begin{bmatrix} 0 & 1 & 1 & 0 \\ -1 & -1 & 0 & 0 \\ 0 & 0 & -1 & -1 \end{bmatrix}$$

We recall that the non identity part of the fundamental cycle matrix, $\overline{F}(D)$, is:

$$\overline{F}(D) = \begin{bmatrix} 0 & -1 & -1 & 0 \\ 1 & 1 & 0 & 0 \\ 0 & 0 & 1 & 1 \end{bmatrix}$$

Then we have that:

$$\bar{K}(D)^{T} + \bar{F}(D) = \begin{bmatrix} 0 & -1 & -1 & 0 \\ -1 & 1 & 0 & 0 \\ 0 & 0 & -1 & -1 \end{bmatrix} + \begin{bmatrix} 0 & -1 & -1 & 0 \\ 1 & 1 & 0 & 0 \\ 0 & 0 & 1 & 1 \end{bmatrix}$$

$$= \begin{bmatrix} 0 & 0 & 0 & 0 \\ 0 & 0 & 0 & 0 \\ 0 & 0 & 0 & 0 \end{bmatrix}$$

It turns out that this pair of matrices will sum to the zero matrix for any digraph. We state this result in the form of a theorem.

THEOREM 8.3: For any digraph, D with fundamental cycle matrix F(D) and fundamental cut-set, K(D), these matrices can be partitioned as:

$$F(D) = [\,I \quad : \bar{F}(D)\,]$$
$$F(D) = [\,\bar{K}(D) : I\,]\ ,$$
$$\bar{F}(D) + K(D)^{T} = 0$$

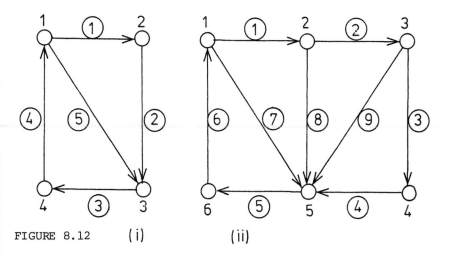

FIGURE 8.12   (i)   (ii)

The reader should follow through the steps below for each of the digraphs in Figure 8.12.

(a)   Find the incidence matrix I(D).
(b)   Find the cycle matrix C(D).
(c)   Show that $C(D)\,I(D)^{T} = 0$.
(d)   Construct a spanning tree of D disorientated.
(e)   Construct the corresponding F(D).

(f) Construct all the cycles of D as linear combinations
     of rows of F(D).
(g) Partition F(D) into $[I : \bar{F}(D)]$.
(h) Construct the fundamental cut-set matrix K(D) for
     this spanning tree.
(i) Generate all possible cut-sets by forming linear
     combinations of the rows of K(D).
(j) Partition K(D) into $\underline{K(D)}_T : I$.
(k) Verify that $\bar{F}(D) + \cdot \bar{K}(D)^T = 0$.
(l) Choose another spanning tree and repeat (h) and (i)
     with this tree.

The key points of the analysis for Figure 8.12(i) are
given in the Solutions to the Exercises, using the
spanning tree whose lines are 2, 4 and 5.

CHAPTER NINE

GRAPHS

9.1 The relationship between graphs and digraphs

In Chapter 8 we mentioned that certain digraphs have the property of being symmetric. This property can be defined as:

$$(v_1, v_2) \in A \Rightarrow (v_2, v_1) \in A, \quad v_1, v_2 \in V.$$

As arcs in symmetric digraphs occur in pairs of the form: (v_i, v_j), (v_j, v_i), mathematicians sometimes combine the arcs of each pair to form one single line or edge which does not have a direction. An example of this arc combination to produce a graph from a digraph is shown in Figure 9.1.

The terms edge and occasionally branch are sometimes used instead of line.

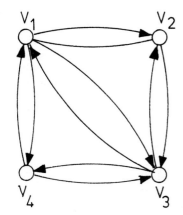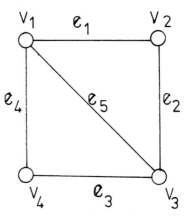

FIGURE 9.1.

DEFINITION 9.1 A graph G = (V,E) is a finite non-empty set V of points, together with a set E of unordered pairs of distinct points of V, called lines.
As an example, the graph in Figure 9.1 has a set of points:

$$V = \{v_1, v_2, v_3, v_4\},$$

and a set of lines:

$$E = \{e_1, e_2, e_3, e_4, e_5\}$$

where

$$e_1 = \{v_1, v_2\},$$

$$e_2 = \{v_2, v_3\},$$

$$e_3 = \{v_3, v_4\},$$

$$e_4 = \{v_4, v_1\},$$

$$e_5 = \{v_1, v_3\}.$$

Note that we use the notation (v_i, v_j) for arcs of digraphs and $\{v_i, v_j\}$ for lines of graphs.   In general

$$(v_i, v_j) \neq (v_j, v_i),$$

as an arc (v_i, v_j) has a specific direction from v_i to v_j.
But a line $\{v_i, v_j\}$ has no orientation and could equally well have been written as $\{v_j, v_i\}$.

A line, e = $\{v_i, v_j\}$, is said to join points v_i and v_j; v_i and v_j are adjacent points, also line is incident with both v_i and v_j; Two lines e_1 and e_2 are said to be adjacent if they are of the form:

$$e_1 = \{v_i, v_j\},$$

$$e_2 = \{v_j, v_k\},$$

i.e. they are both incident with some point v_j.

In a digraph the number of precursors of a point p (i.e. the number of arcs ending at p) is its indegree, id(p).
Also the number of successors of a point p (i.e. the number of arcs beginning at p) is its outdegree, od(p). When a graph is constructed from a digraph each line created accounts for two arcs: one directed toward and one away from a specific point. Hence the number of lines incident with a point p, d(p), obeys the relationships:

$$d(p) = od(p) = id(p).$$

DEFINITION 9.2 The degree d(p) of a point (p) in a graph (V,E) is the number of lines incident with p.

THEOREM 9.1: The sum of the degrees of the points in a graph is equal to twice the number of lines in the graph.

PROOF   This can be proved via the relationship between the
graph and the corresponding symmetric digraph, and Theorem
1.1.
    A direct proof is to observe that summing the degrees
sums the ends of lines, while each line has two ends.   □

9.2 Subgraphs and isomorphism

    The concepts of subdigraph and partial digraph (Section
1.11) and isomorphism (Section 1.13) have direct parallels
in the theory of graphs.

> DEFINITION 9.3   If $G = (V,E)$ is a graph, let U be a
> non-empty subset of V.   Then U defines a subgraph of
> G having U as its point set and all the lines of G
> which have both endpoints in U as its set of lines.
>   For example, let G be the graph in Figure 9.1(b), and
> let $U = \{v_1, v_2, v_3\}$.   Then the line set of the subgraph is
> $\{e_1, e_2, e_5\}$.
>   Reversing the roles, if (U,F) is a subgraph of (V,E)
> we also describe (V,E) as a supergraph of (U,F).

> DEFINITION 9.4   If $G = (V,E)$ is a graph and F is a
> subset of E, then we say that (V,F) is a partial
> graph of (V,E).
>   In drawing a graph or a digraph the relative positions
> of the points and the layout of the lines or arcs, subject
> to the constraint that they remain linked to the correct
> points, have no significance: the point, line and arc sets
> carry all the information.   Graphs which can be transformed
> one into the other by deformation, as though the points
> were small blocks joined by rubber bands to represent the
> lines and capable of sliding over the page, together with
> relabelling of points, are called isomorphic.   The formal
> expresssion of this relationship appears somewhat more
> formidable.

> DEFINITION 9.5   Let $G_1 = (V_1, E_1)$ and $G_2 = (V_2, E_2)$;
> then an isomorphism from G_1 to G_2 is an invertible
> function f from V_1 onto V_2 (i.e. a one-one
> correspondence) such that if $f(u_i) = v_k$ and $f(u_j) = v_m$,
> then v_k and v_m are adjacent if and only if u_i and
> u_j are adjacent.   Symbolically we may write
>
> $$E_2 = \{\{f(u_i), f(u_j)\} : \{u_i, u_j\} \in E_1\}.$$

We say that G_1 and G_2 are <u>isomorphic</u> if an isomorphism from
G_1 to G_2 exists, and write $G_1 \cong G_2$.

Isomorphism can be shown to be an equivalence relation,
partitioning the set of all graphs.

The graphs in Figure 9.2 are isomorphic, for if we use
the one-one correspondence

$$f(u_i) = v_i,$$

we see that adjacency is preserved, the two line sets being
$\{\{u_1,u_4\},\{u_1,u_5\},\{u_1,u_6\},\{u_2,u_4\},\{u_2,u_5\},\{u_2,u_6\},\{u_3,u_5\},$
$\{u_3,u_6\}\}$, and
$\{\{v_1,v_4\},\{v_1,v_5\},\{v_1,v_6\},\{v_2,v_4\},\{v_2,v_5\},\{v_2,v_6\},\{v_3,v_5\},$
$\{v_3,v_6\}\}$ where the lines have been laid out to display the
correspondence.

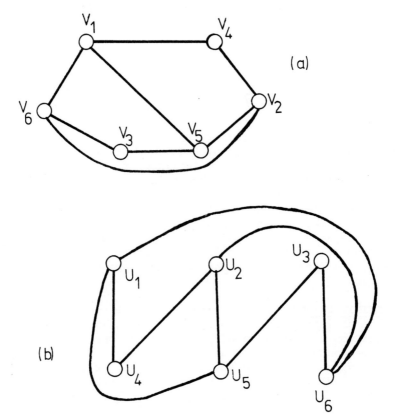

FIGURE 9.2   Isomorphic graphs

9.3 Complete, Complementary and Self-complementary Graphs

Consider the graph in Figure 9.3. You will notice
that it has every possible line - every pair of points are
adjacent. A graph with n points, every pair being adjacent
is termed complete and is denoted by K_n;   as each
point p is directly connected to every one of the other
(n-1) points it has degree
$$d(p) = n-1$$
Also as there are n points, each incident with (n-1) lines,
K_n has
$$\frac{n(n-1)}{2} = \binom{n}{2}$$
lines. The expression on the left is divided by 2 as we
have counted each line twice. The graph K_3 is called a
triangle for obvious reasons.

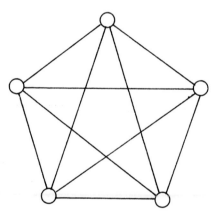

FIGURE 9.3 The complete graph on 5 points

Consider the graph, G in Figure 9.4(a). Let us create
a graph, G' with the same point set, V, but which has a
line connecting two points if and only if the two are not
connected in G. Then G' is shown in Figure 9.4(b). When
a graph G' is constructed in this way from a graph G it is
said to be the complement of G, and the two graphs are
complementary. A moment's reflection will provide a
proof of the following theorem, which states that the
complement of the complement of a graph is the graph
itself.

THEOREM 9.2   For any graph, G
$$(G')' = G$$

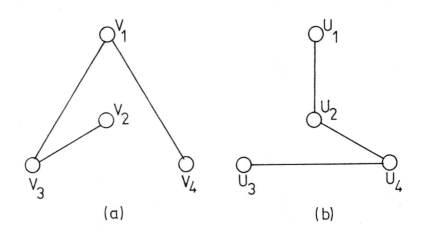

FIGURE 9.4   Complementary graphs

Let us investigate the two graphs in Figure 9.4 a little further. Suppose we define the following 1-1 correspondence relation, f between the points of the two graphs according to how they are labelled in the figure:

$$f(v_1) = u_2,$$
$$f(v_4) = u_3,$$
$$f(v_3) = u_1,$$
$$f(v_2) = u_3.$$

Then it can be seen that a preserves adjacency and hence G' is isomorphic to G. When a graph is isomorphic to its complement it is said to be <u>self-complementary</u>. Two isomorphic digraphs must clearly have the same number of lines, and between them they have precisely the lines of the complete graph, so they must have half each. The number of lines in a self-complementary graph is thus

$$\tfrac{1}{4}n(n-1).$$

As this must be an integer, but n and n-1 are not both even, then either n is a multiple of 4 or n is one more than a multiple of 4. The smallest possible values for n are thus   1,4,5,8,9,12,13,...
and it can be shown that there are self-complementary graphs for each of these numbers.

We have now developed the tools to analyse a little

puzzle. Recall that in Chapter 7 we analysed the project
of planning a dinner party. Suppose that six people are
present at the party. How can one prove the statement that
    (a) Either there are three people who all know each
        other
    (b) Or there are three people who do not know each
        other?
    To solve this problem we create a graph, G with six
points, one for each person. If two people are acquainted
the appropriate line is drawn in the graph. We now create
a second graph, G' with the same points. If two people are
not acquainted the appropriate line is drawn in the graph.
As each pair is either acquainted or not, these two graphs
are complementary and together they produce K_6. Now if
three people all know each other there will be a triangle
of edges in G. If three people do not know each other
there will be a triangle of edges in G'. Hence the problem
reduces to proving for any two complementary graphs G and
G' on six points that either G contains K_3 or G' contains
K_3. To prove this consider one person, p. Now as each
other person is considered in turn, p either knows him or
he does not. There are five other people. Hence point p
is incident with at least three lines in either G or G'.
Let us assume it is in G'. (The argument is the same for
G.) Let the three points, other than p, with which these
lines are incident by v_1, v_2, and v_3. Consider the three
people: v_1, v_2, and v_3.
If they all know each other their points form a triangle in
G. If two of them do not know each other, their points will
be connected by a line in G'. This line forms part of a
triangle with p in G' and the proof is complete.

9.4 Exercises

    1. Prove that if G_1 and G_2 are two isomorphic graphs, then
       (a)   they have the same number of lines,
       (b)   they have the same number of points of each degree
          degree.
    2. Prove that graph isomorphism is an equivalence relation.
S 3. Figure 9.5 shows the children who attended a party.
       Two names are joined by a line if the children belong
       to the same family or were in the same class at school.
       Given that Rebecca, Timothy and Stephen form one
       family, assign all lines to two sets, one for 'same
       family' and one for 'same class'. What have complete
       subgraphs to do with this classification? (No two

children in the same family were also in the same class).

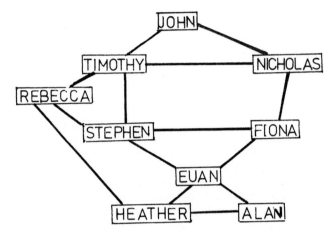

FIGURE 9.5

S 4. At Canterbury University students may choose from seven first-year mathematics courses, as in the table below.

Course no.	Course name	Credit pts.	Restriction
102	Elementary Calculus	4	110,112
103	Elementary Finite Mathematics	4	110,111
104	Elementary Abstract Mathematics	4	111
110	Calculus and Linear Algebra	6	102,103,111, 112
111	Algebra	6	103,104,110
112	Calculus	6	102,112
113	Mathematical Methods	6	

Students are not permitted to count to their degree combinations involving both any course and any course in its restriction column.
Draw a digraph showing the restrictions and observe that it is symmetric. Draw the corresponding graph. Draw the complement. What graph-theoretic term can be

applied to the subgraph of this complement defined by any allowed combination of courses? Find an allowed combination carrying the maximum number of credit points.

S 5. Given the adjacency matrices of the following digraphs draw the associated graph for each using arc combination.

$$\begin{bmatrix} 0 & 1 & 1 & 1 \\ 1 & 0 & 1 & 1 \\ 1 & 1 & 1 & 0 \\ 1 & 1 & 1 & 1 \end{bmatrix} \qquad\qquad \begin{bmatrix} 0 & 1 \\ 1 & 0 \end{bmatrix}$$

(a)   (b)

$$\begin{bmatrix} 0 & 1 & 1 & 0 & 0 & 0 \\ 1 & 0 & 1 & 0 & 0 & 0 \\ 1 & 1 & 0 & 0 & 0 & 0 \\ 0 & 0 & 0 & 0 & 1 & 1 \\ 0 & 0 & 0 & 1 & 0 & 1 \\ 0 & 0 & 0 & 1 & 1 & 0 \end{bmatrix} \qquad \begin{bmatrix} 0 & 1 & 0 & 0 & 0 & 0 & 0 \\ 1 & 0 & 1 & 0 & 0 & 0 & 0 \\ 0 & 1 & 0 & 1 & 0 & 0 & 0 \\ 0 & 0 & 1 & 0 & 1 & 0 & 0 \\ 0 & 0 & 0 & 1 & 0 & 1 & 0 \\ 0 & 0 & 0 & 0 & 1 & 0 & 1 \\ 0 & 0 & 0 & 0 & 0 & 1 & 0 \end{bmatrix}$$

(c)   (d)

6. Let n be a positive integer. Consider the integers from 1 to $n-1$. Call a pair $\{a,b\}$ of such integers a 'factorisation of zero' if $a.b$ is a multiple of n. (Students who know about congruence of integers should understand the strange terminology, and also be able to recast the definition). For any such n we may form a graph whose points are 1 to $n-1$ and whose lines are the factorisations of zero. Construct the graphs for $n=6$, $n=10$ and $n=30$. What happens if n is prime; what happens if it is divisible by a square?

7. Consider the set of all English words of four letters. Take this as the point set of a graph whose lines join pairs of words which differ in just one position, such as {host, hoot}, {hoot, foot}, {post, port}. Construct a large subgraph of this graph.

S 8. If G_2 is a subgraph of a graph G_1, what is the relationship between their adjacency matrices?

S 9. If G_2 is a partial graph of a graph G_1, what is the relationship between their adjacency matrices?

S 10. Let G be a graph. A subgraph H of G is called a
      <u>clique</u> if (a) H is a complete graph and (b) there
      is no complete graph K which is both a subgraph of G
      and a supergraph of H (other than H itself). Find the
      cliques of the graph in Figure 9.6.

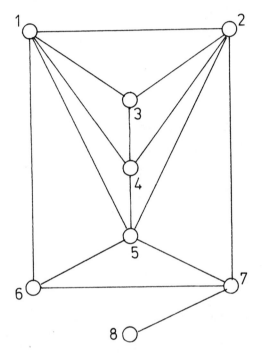

FIGURE 9.6

11. Given any graphs $G_1, G_2, \ldots, G_m$, with $p_1, p_2, \ldots, p_m$
    points respectively, show that there is a graph G with
    subgraphs isomorphic to each of $G_1, \ldots, G_m$. Try to
    reduce the number of points by overlapping subgraphs.
    (A complete and exact general minimum is not to be
    expected).

12. If a drama group is short of actors for a particular
    play it may 'double up' parts: that is, one actor
    plays several parts. Naturally he cannot play two
    parts that require him to be on stage together. To
    each play corresponds a graph, whose points are the
    characters and whose lines join characters on stage
    together at some point in the play. How does a set of

parts which can be doubled up appear in the
complement of this graph? Take a play and (by
cooperative effort) draw up its graph and complement.
Consider the scope for doubling up.

S 13. State with reasons which of the graphs in Figure 9.7
are self-complementary.

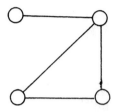

a

b

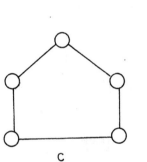

c

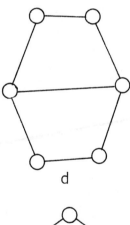

d

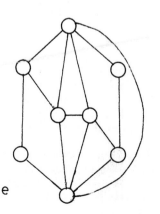

e

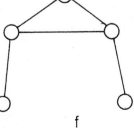

f

FIGURE 9.7

14. Which of the graphs in Figure 9.7 are isomorphic to
    subgraphs of other graphs in the same figure? Give
    your reasons.
15. Construct a graph which has subgraphs isomorphic to
    each of the graphs in Figure 9.7. (It can be done
    with nine points).
16. Draw a graph whose points represent the rooms of your
    house, or some other building you know, and whose
    lines indicate that there is a doorway between the two
    rooms. Can your graph be drawn so that no two lines
    cross? (Whether it can or not will depend on the
    design of the house).

9.5   Planar and Bipartite graphs

Consider the graph in Figure 9.2(b). Suppose that
u_1, u_2, and u_3 represent houses. Across the street are
situated the gas company, the electric company and the
telephone company, represented by u_4, u_5, and u_6
respectively. Each line represents a cable or pipe
bringing one of the services to one of the houses. You can
see that each house is supplied with each service except
that u_3 cannot have a gas stove: the $\{u_3, u_4\}$ line is
missing. Now suppose that city bylaws prevent cables from
overlapping and the owners of the other houses or companies
will not allow a cable to be passed through their premises.
Is it possible to layout the cables in another way so that
u_3 can have the gas hooked up?

To save you hours of time we tell you now it isn't
possible. Now this schoolboys' puzzle introduces the
concept of planarity. Loosely speaking a planar graph is
one which can be drawn in the plane without line
intersection. So the graphs in Figure 9.2 are planar. Now
look at the graph in Figure 9.8. It attempts to solve
the previous problem by simply laying out the cables as
straight lines. Now as we said that there was no solution
to the puzzle, this means that this graph is nonplanar in
the sense that it cannot be drawn in the plane without line
intersection. The graph K_5 in Figure 9.3 is also
nonplanar and is the only nonplanar graph with less than
six points.

Much attention has been paid by graph theorists to
planar graphs. In particular they are connected with the
famous four colour problem, solved by Haken and Appel in
1976. Many of the books mentioned under 'Further Reading'
devote a chapter or more to planar graphs.

Let us look at the graph in Figure 9.8 more closely.

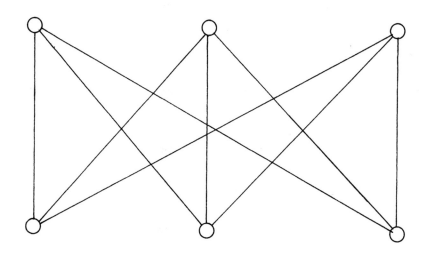

FIGURE 9.8 The complete bipartite graph $K_{3:3}$

Recall that we partitioned the set of points, V into two
subsets: U (= $\{u_1, u_2, u_3\}$ the set of houses), and
W (= $\{u_4, u_5, u_6\}$, the set of companies). No lines are
inserted between points in the same set. Hence all lines
connected points of different sets. Graphs which obey
this definition are said to be <u>bipartite.</u> The graph in
Figure 9.8, which has three points in each set is denoted
by $K_{3:3}$. In general when the two subsets of points have
n and m points and every one of the n points is joined to
every one of the m points then the bipartite graph is
denoted by $K_{n:m}$. When all possible lines joining the two
sets are present, the graph is called a <u>complete</u>
<u>bipartite</u> graph. Of course a bipartite graph need not have
all possible lines present - the operative part of the
definition is the sentence: no lines are inserted between
points of the same set. An example of a bipartite graph
which is not complete is a polygon with an even number of
points. To see this start at any point, proceed round the
polygon labelling the points: 1, 2, 3, ... The even
numbered points comprise one set, the odd numbered points
the other.

9.6   <u>Trees</u>

The concepts of <u>walk</u>, <u>path</u>, <u>cycle</u> and <u>distance</u> can be
applied to graphs in a way analogous to these concepts in

digraphs, the correspondence between graphs and symmetric
digraphs providing the link. For convenience we state them
again in graph terms: notice that there is a significant
difference in the definition of cycle.

DEFINITION 9.6   A walk is a sequence of points and
lines alternately, beginning and ending with points,
and such that each line is incident with the points
written before and after it, and these two points must
be different.
A walk is closed if its first and last points coincide.
A path is a walk in which no point occurs more than
once.
A cycle is a closed walk in which no point is
repeated, except the first and last point, and in
which there are three distinct points.

The last part of the definition of a walk prevents
such sequences as
            u, {u,v}, v, {v,w}, v, {v,x}, x.
The restriction on the number of points in a cycle
excludes both the trivial cycle consisting of a single
point, and also traversing a single line in both directions
directions:
            u, {u,v}, v, {v,u}, u.
Although lines are undirected, each walk has a
direction defined by the way in which it is written.
Because of the undirected nature of lines, each walk is
reversible, so that reachability becomes a symmetric
relation. Hence the inreach, outreach and component of a
point in a graph coincide, and we write of components.
Strong and weak connectedness also coincide.

DEFINITION 9.7   A graph is connected if given any two
points there is a walk from one to the other.   A
component of a graph is a maximal connected subgraph.

DEFINITION 9.8 The length of a walk is the number of
lines it contains. The distance between two points
is the length of the shortest walk between them.   The
diameter of a connected graph is the maximum distance
between any two points.

In Figure 9.7 observe that all graphs are connected,
and that all except (b) have cycles. Connected graphs
without cycles are important in graph theory.

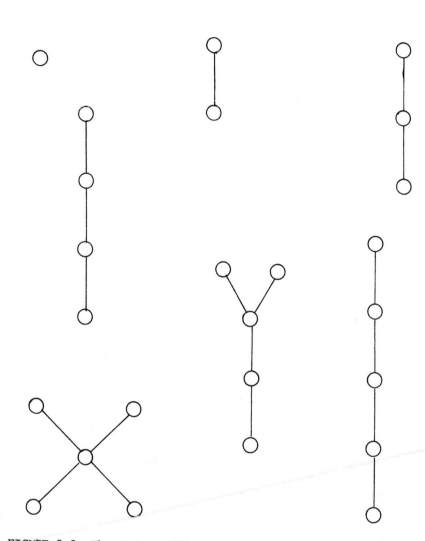

FIGURE 9.9   The trees with at most five points

DEFINITION 9.9   A graph without cycles is termed a
forest.   A connected graph without cycles is called a
tree.

In Figure 9.9 are shown all trees (to within
isomorphism) with at most five points.   Trees arise in
several contexts.   In particular a tree is obtained
whenever a rooted tree is disoriented (Section 8.6).
Conversely if any point of a tree is chosen as a root, and

every line is replaced by an arc directed away from the
root (which direction is uniquely defined), a rooted tree
results. A sketch of a botanical tree, or any plant, will
also yield a tree - with a few exceptions. It is the same
with the tributaries of a river: again there are
exceptions. Nature is never quite as tidy as we would like.

There are a number of equivalent definitions of a
tree. Each of the following five statements defines
exactly the same set of trees.

(a)   a graph with n points and n-1 lines, and without
      cycles;

(b)   a graph with n points and n-1 lines, and
      connected;

(c)   a graph with exactly one path between each pair
      of points;

(d)   a graph which is connected and without cycles;

(e)   a minimally connected graph. (A graph is
      minimally connected if it is connected and
      removal of any line creates a non-connected graph)

Note that (d) is our original definition of a tree.

Before proving the equivalence of these statements we
prove a Lemma.

LEMMA 9.3: Let $G = (V,E)$ be a connected graph with n
points and at least n lines. Then G contains a cycle.

PROOF We construct sets $X_1, X_2, \ldots, X_n = V$ in the following
way. Let u_1 be any point of G. Let $X_1 = \{u_1\}$. Then for
$i = 1, 2, \ldots, n-1$

$$X_{i+1} = X_i \cup \{u_{i+1}\},$$

where u_{i+1} is a point chosen in the following manner.
Suppose X_i has been constructed, and $i < n$. Then
there is at least one point v_i not in X_i. As G is
connected there is a walk from u_1 to v_i; Following this
walk we must come to some first point not in X_i. Call
this point u_{i+1}, and let the line which precedes it in the
walk from u_1 to v_i be e_i. Thus finally $X_n = V$, and a set
$H = \{e_1, e_2, \ldots, e_{n-1}\}$ has also been constructed. As there
are at least n lines altogether there exists a line f which
is in E but not in H. Let $f = \{u_i, u_j\}$. Then there are
walks of lines in H from u_1 to u_i and from u_1 to u_j.
Reversing the former and joining them we obtain a walk from
u_i to u_j, and from this can be extracted a path (as in
Section 1.8), involving at least one other point besides
u_i and u_j, and this path together with f forms a cycle. □

In all the following theorems assume that G is a graph
with n points.

THEOREM 9.4

$$(a) \Rightarrow (b)$$

PROOF: Assume G has (n-1) lines and no cycles. We shall
also assume that G is disconnected and show that this is
impossible. Without loss of generality suppose that G
comprises two components: C_1 and C_2. As G has n-1 lines,
one component must have at least as many lines as points.
But this component must then contain a cycle by the above
Lemma.   □

THEOREM 9.5

$$(b) \Rightarrow (c)$$

PROOF Assume G has n-1 lines and is connected. We shall
also assume that G has more than one path between a pair of
points, p_1 and p_2 and show that this is impossible. The
existence of at least two paths between p_1 and p_2 will
create a cycle. Suppose this cycle is incident with m
points. Hence it comprises m lines, and

$$m < n$$

as G has only n-1 lines.   It has taken m lines to
connect together the m points of the cycle. To connect a
further point to the cycle must use up a further line.
Thus in the process of connecting the extra points to the
cycle by existing lines, eventually n-1 points will be
connected and all the n-1 lines will be used up. Thus the
nth point will not be connected to any of the others.   □

THEOREM 9.6

$$(c) \Rightarrow (d)$$

PROOF Assume G has exactly one path between each pair of
its points. Then G is certainly connected. Further G
cannot have any cycles if there is to be exactly one path
between each pair of points.   □

THEOREM 9.7

$$(d) \Rightarrow (e)$$

PROOF Assume G is connected without cycles. We shall also
assume that G is not minimally connected and show that this
is impossible. If G is not minimally connected then there
exists at least one line, e which can be removed without
disconnecting its endpoints, p_1 and p_2. Now the path still
connecting p_1 and p_2, together with e forms a cycle.   □

THEOREM 9.8

$$(e) \Rightarrow (a)$$

PROOF   Assume that G is minimally connected. The process
used in the proof of Lemma 9.3 in fact shows that for any
connected graph (V,E) there is a connected partial graph
(V,H) which has exactly (n-1) lines and that no set of
fewer lines will do. As G is minimally connected we must
have E = H. Hence G has exactly (n-1) lines.

        If G contains a cycle, the deletion of some line does
not destroy the connectivity. Hence no graph with a cycle
can be minimally connected.

        Thus a minimally connected graph with n points has
(n-1) lines and no cycles.   □

        The previous five theorems have proven that the
definitions (a) - (e) all equivalently define a tree.

        Look at the graphs with at least two points in Figure
9.9. You will notice that in each there are always at least
two points "dangling" by only one line. These points, p
which can be characterized by the equation:

$$d(p) = 1$$

are called pendant points. The fact that there are at least
two pendant points in each graph is no coincidence. For in
any tree of n points there are n-1 lines. Each line
contributes 2 to the sum of the degrees. So the sum of the
degrees of all the points is 2(n-1). Now because a tree is
connected each point must have degree at least one. But
even if the degrees are distributed as evenly as possible,
and each point gets no more than two, there will still not
be enough to go round and at least two points must have
degree one. We have proven:

THEOREM 9.9   In any tree with at least two points there are
at least two pendant points.

        The set of lines H constructed in any connected
graph G in Lemma 9.3 is, by the application of alternative
description (b) of a tree, the line set of a tree.

        It is also a partial graph of G. A partial graph of a
graph which is a tree is called a spanning tree of the
graph. It is quite easy to find a spanning tree of a
connected graph. One simply removes lines at random from
the graph, taking care not to disconnect the graph. When
no further lines can be removed, what is left is a spanning
tree. This has been done for the graph in Figure 9.10,
the lines $c_1, c_2, \ldots, c_5$ having been removed.

        Lines of a spanning tree, T of a graph are called
branches, the other lines of the graph are called chords
with respect to T. (See also p.189).

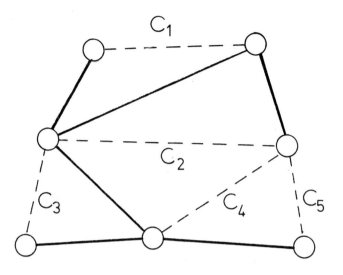

FIGURE 9.10   A spanning tree of a graph

Now suppose G has n points and m lines. Now we know
that a tree on n points has n-1 lines. Hence there will be
n-1 branches and m - (n-1) chords. The latter part of
Chapter 8 discusses some concepts concerned with spanning
trees in digraphs. All these concepts carry over to graphs.
    In many applications of graph theory it is convenient
to associate a real number or weight with each line of a
graph, just as we did with the arcs of a digraph in some
applications in Chapters 5 and 6. The weight is often a
cost or gain in terms of time, money, distance or some
other criterion. As an example suppose an emerging
country is considering developing a railway system to
connect its cities. The cost of linking every pair of
cities is known. The president wants to make sure travel is
possible between every pair and the whole system must be
built at minimal cost. Now if we represent cities by
points and railway track by lines we wish to find the
spanning tree with the smallest total line weight. This
spanning tree is called the minimal spanning tree. A
minimal spanning tree for any weighted graph can be

constructed by <u>Kruskal's Method</u>. This will be explained
by an example, using Figure 9.11, in which the circled
numbers represent the weights.

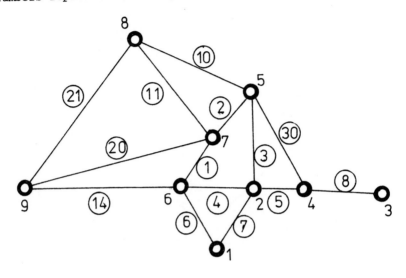

FIGURE 9.11   A graph with distances

     Kruskal's method begins by ordering the lines in non-
decreasing order of weight. Hence the first line is {6,7}
with weight 1.   Then {5,7} with weight 2, then {2,5},{2,6},
{2,4}, {1,6}, {1,2}, {3,4}, {5,8}, {7,8}, {6,9}, {7,9},
{8,9}, and {4,5}. Now the first line, {6,7} is selected to
become part of the tree. Each line is considered in turn,
and is selected as part of the tree unless it makes a cycle
with lines already selected, in which case it is rejected.
So lines {5,7} and {2,5} are selected. The next line is
{2,6}, which is rejected because its selection would
create the cycle:
     <6, {6,7}, 7, {7,5}, 5, {5,2}, 2, {2,6}, 6 >
So we pass on to {2,4} which is selected, then {1,6} which is
selected. Line {1,2} is rejected but {3,4} and {5,8} are
selected and {7,8} is rejected. Line {9,6} is selected and
the last three are rejected. This creates the minimal
spanning tree shown in Figure 9.12, which has a total
weight of 49.
     That Kruskal's algorithm works can be seen by the
following argument. Let K be the set of lines chosen by
Kruskal's algorithm, and let L be the set of lines in any

other spanning tree of the graph.   Both K and L contain
(n-1) lines.   If K $\neq$ L there is a first line k $\in$ K in the
ordering by which K is defined which is not an element of
L.

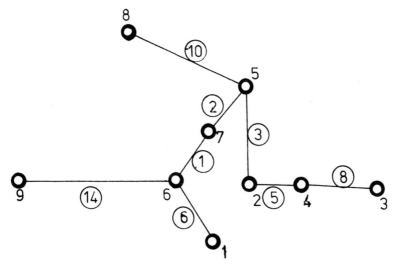

FIGURE 9.12   The minimal spanning tree of Figure 9.11

    If we add k to L the resulting graph has a cycle C,
including k.   Not every line of C is in K, for if they were,
K would not define a spanning tree.   If the lines of C are
arranged in order of weight, as for the construction of K,
the last line has the highest weight.   Call this line e.
As k is the least weight line of K not in L, all lines of C
shorter than k are in L.   Then k = e would mean that K
contained a cycle.   Hence k occurs before e.   If then a new
spanning set M is formed from L by deleting e and adding k,
the total weight of M is less than that of L, unless e and
k had the same weight, when they would be equal.
    Now comparing M and K another line of M can be
replaced by a line of K, and so by a sequence of trees
whose total weight is at worst not increasing, we
eventually reach K.   Thus K is a spanning tree (not
necessarily the only one) of a minimal total weight.   In
fact all minimal total weight spanning trees are obtained
by the Kruskal method for some ordering of lines of equal
weight.

9.7 Exercises

1. Airlines often publish graphs of the routes they fly.
   Get hold of such a graph. Decide whether or not it is
   planar. (We have not given you the information to
   enable you to <u>prove</u> whether it is plaar or not).

S 2. Which of the graphs in Figure 9.7 are bipartite? Give
   an assignment of the points to the two sets for those
   that are, and a proof that the assignment is
   impossible for those that are not.

S 3. Prove that every tree is bipartite.

4. Draw all the complete bipartite $K_{m:n}$ where $m + n \leqslant 6$.

S.5 Describe all graphs which are both trees and complete
   bipartite.

S 6. Construct a definition for a <u>centre</u> of a graph, on the
   lines of Definition 2.11. Prove that a tree has
   either one centre or two adjacent centres.

S 7. Draw the planar graphs in Figure 9.13 in such a way
   that no lines intersect.

S*8. Suggest a practical application of bipartite graphs.

*9. Suggest a practical application of trees.

10. The concept of a graph complement can be generalised.
    One way is to define the sum of two graphs on the
    same set of points. Suppose $G_1 = (V,E_1)$ and
    $G_2 = (V,E_2)$, and that $E_1 \cap E_2 = \emptyset$. Then define

    $$G_1 + G_2 = (V, E_1 \cup E_2).$$

    Then two graphs are complementary if $G_1 + G_2$ is
    defined and is the complete graph on V. Look for
    solutions to the equations.
    (a) $T_1 + T_2 = K_n$, where T_1 and T_2 are trees;
    (b) $T_1 + T_2 = K_{m:n}$, where T_1 and T_2 are trees;

S 11. A <u>cactus</u> is a connected graph in which no line is in
    more than one cycle: a <u>smooth cactus</u> is a connected
    graph in which every line is in precisely one cycle.
    Which graphs in Figure 9.7 are cacti? Which are
    smooth cacti? Draw some cacti with eight points.

S 12. Let C be a smooth cactus with k cycles (i.e. only
    counting cycles which are not obtained by simply
    changing the starting point), lengths $c_1, c_2, \ldots, c_k$.
    How many lines has C? How many points has C?

S 13. Prove that if G is a non-connected graph then its
    complement G' is connected, and has diameter at most
    2.

*14. Prove that if G is a connected graph with diameter 4
    or more, then its complement G' is connected and has
    diameter 2.

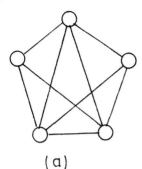

(a)

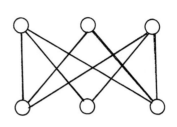

(b)

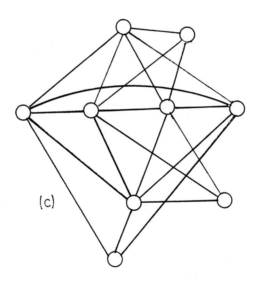

(c)

FIGURE 9.13

*15. Prove that a connected graph cannot be isomorphic to
    a non-connected graph.
*16. Prove that if G and H are both connected graphs, and
    are isomorphic, then their diameters are equal.
17. Use Exercises 13, 14, 15 and 16 to prove that a self-
    complementary graph is necessarily connected, and has
    diameter either 2 or 3.
18. Prove that every tree is planar.
S 19. State and prove a simple rule for constructing the
    minimal spanning tree of a cactus (Exercises 11 and 12)

S 20. In any graph, let v_k be the number of points of degree
k.   For each of the sets of values (a) to (g),
decide whether such a graph is possible; whether it
must be, or can be, a tree; whether it must be, or can
be, bipartite; whether it must be, or can be,
connected; whether it must be, or can be, self-
complementary.   In each case there is no point of
degree greater than 6.   List your rules for deciding.
Prove some of them.

	v_0	v_1	v_2	v_3	v_4	v_5	v_6
(a)	0	0	5	0	0	2	0
(b)	0	1	3	2	0	0	0
(c)	0	8	0	1	1	1	0
(d)	0	7	1	1	0	0	0
(e)	0	0	3	6	0	0	0
(f)	0	2	0	2	2	0	2
(g)	5	4	3	2	1	0	0

21. State a further definition of a tree.

S 22. Find minimal spanning trees for the graphs in
Figure 9.14 using Kruskal's method.

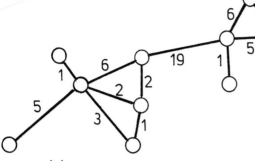

(a)

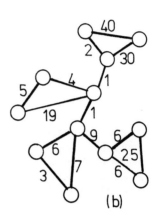

(b)

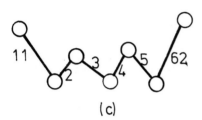

(c)

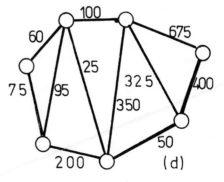

(d)

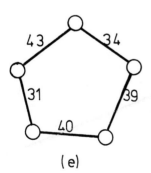

(e)

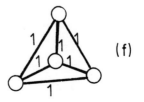

(f)

FIGURE 9.14

SOLUTIONS TO SELECTED PROBLEMS

CHAPTER 1.

Exercises 1.3

1. The equivalence classes are {stop, post, pots, spot, tops}, {sport, ports}, {spots, posts}, {top, pot}, {port}, {sort}. In the arrow diagram there is a loop at each word and an arrow from each word to each other word to each other word in the same class.

2. The relation is not transitive as 'post' is related to 'port', and 'port' to 'sort', but not 'post' to 'sort'. The relation is also not reflexive. It is antireflexive and therefore is a digraph.

5. The pairs are (1,A), (1,D), (2,C), (2,D), (3,C), (3,D), (4,A), (4,B), (4,C), (4,D), (5,A), (5,B), (5,D) for flowers and bees. For the second relation the pairs are (1,1), (2,2), (3,3), (4,4), (5,5), (1,4), (1,5), (2,3), (2,4), (3,2), (3,4), (5,4).

Exercises 1.6

1. $V = \{1,2,3,4,5,6\}$, $A = \{(1,2), (1,3), (1,6), (2,3), (3,4), (4,5), (5,6), (6,4)\}$.

2. $V = \{1,2,3,4,5,6,7\}$, $A = \{(1,2), (1,6), (1,7), (2,3), (3,7), (4,3), (4,5), (5,6), (5,7), (6,4), (7,2), (7,4), (7,6)\}$.

3.

Vertex	precursors	successors	indegree	outdegree
1	-	2,6,7	0	3
2	1,7	3	2	1
3	2,4	7	2	1
etc.				

5. The number of buffers equals the number of ramps plus
    the number of branches.

Exercises 1.10

1. x,(x,e),e,(e,f),f,(f,h),h,(h,x),x,(x,y),y,(y,e),e;
    (x,e),(e,f),(f,h),(h,x),(x,y),(y,e).
    It is not a path, nor is it closed.

2. Spanning walks: A,B,C,E,G; spanning paths: A,B,C,E,G;
    spanning closed walks: B,G; spanning cycles: None.

3. Delete one arc from the spanning cycle.

6. p-1

7. There are two shortest cycles, of length 9.
    One is $<$ PB,13,8,4,3,2,6,7,12,PB $>$.

10.

	t	m	e	c	r
t	0	1	2	3	3
m	1	0	1	2	2
e	1	1	0	1	1
c	1	2	1	0	2
r	1	2	2	1	0

12. One definition is: if $W = <w_1, w_2, \ldots, w_k>$ and
    $X = <x_1, x_2, \ldots, x_m>$ are walks in a digraph, then
    $W = X$ if and only if $k = m$ and $w_i = x_i$ for $1 \leq i \leq k$.

13. If C then not A; not C.

14. No well-fed canary is melancholy (i.e. no melancholy
    canary is well-fed).

17. No jubjub is a jubjub. That is, jubjubs do not exist.

Exercises 1.14

8. The avoidable one-way streets are (5,3), and (6,7). If
    the one-way streets are omitted, the remaining partial
    digraph, which is symmetric, falls into three
    components (portions). Avoidable one-way streets join
    points in the same component.

10. The two isomorphisms are:

    $1 \rightarrow b$, $2 \rightarrow d$, $3 \rightarrow c$, $4 \rightarrow e$, $5 \rightarrow a$; $1 \rightarrow c$, $2 \rightarrow e$, $3 \rightarrow b$,
    $4 \rightarrow d$, $5 \rightarrow a$.

11. Because q and 5 are the only sources in the digraphs,
    they must correspond in any isomorphism. But in (X,C)
    one of the successors of q is a precursor of the other,
    while in (V,A), neither is a precursor of the other.

CHAPTER 2

Exercises 2.4

1.
Vertex	Inreach	Outreach
1	{1}	{1,2,3,4,5,6}
2	{1,2}	{2,3,4,5,6}
3	{1,2,3}	{3,4,5,6}
4	{1,2,3,4,5,6}	{4,5,6}
5	{1,2,3,4,5,6}	{4,5,6}
6	{1,2,3,4,5,6}	{4,5,6}

The components are {1}, {2}, {3}, {4,5,6}

3. The set of people each can persuade is his outreach (or
   the person himself can be removed from this set: it
   does not affect the comparison). The most powerful
   person is c: his outreach has 5 members.

4.
Vertex	Inreach
C1	{C1}
LA1	{LA1}
NS1	{NS1}
MP2	{C1, LA1, MP2}
LA2	{LA1, NS1, LA2}
AA2	{LA1. NS1, AA2}
MP3	{C1, LA1, NS1, MP2, LA2, MP3}
LA3	{LA1, NS1, LA2, LA3}
AA3	{LA1, NS1, AA2, AA3}

There are seven valid combinations of five courses.

5. The inreach in the partial digraph is contained in the
   inreach in the original digraph. The same relationship
   holds between the components.

7.
Vertex	Successors	Outreach
1	2,6,7	1,2,6,7,3,4,5
2	3	2,3,4,5,6,7
3	4	3,4,5,6,7,2
4	3,5	4,3,5,6,7,2

Vertex	Successors	Outreach
5	6,7	5,6,7,2,4,3
6	4	6,4,3,5,7,2
7	2,4,6	7,2,4,6,3,5

The components are {1}, {2,3,4,5,6,7}.

9. The components are, using initials, {a,b,c,d}, {e,f,g},
   {h,i,k}, {j}, {m}. Components consist of groups of
   close friends. John and Michael chose friends who were
   in different groups from the people who chose them.

Exercises 2.8

1. The components, and vertices of the condensation, are
   {1}, {2}, {3}, {4,5,6}. Calling these A,B,C,D in
   order, the arcs of the condensation are (A,B), (A,C),
   (B,C), (A,D), (C,D).

2. The components of the converse of a digraph are the
   same as the components of the original digraph. The
   condensation of the converse is the converse of the
   condensation of the original digraph.

3.

Components	Ingates	Outgates
{ECON, POLS}	ECON, POLS	ECON, POLS
{EDUC, ENGL, GEOG, FREN,	EDUC, ENGL, GEOG,	ENGL
GRMN, HIST, PSYC, SOCI}	SOCI	
{JAPA}		JAPA
{MUSI}		MUSI
{AMST}	AMST	
{MATH}	MATH	

5. PROOF   Let $W = \langle u = u_1, u_2, \ldots, u_m = v \rangle$ be any walk
   from u to v in D. Consider consecutive points,
   u_i, u_{i+1}. Either they are both in the same component
   C, or they are in different components, say C_1, C_2. In
   the latter case D* has also an arc (C_1, C_2). Thus
   following W we always either stay at the same point of
   D* or follow an arc of C*. Writing down the
   components as we come to them gives the corresponding
   walk W* in D*. Then C(v) is reachable from C(u) in D*.
       Notice that this proof is in fact wider than we
   asserted: the result applies to any factor digraph,
   not just the condensation. On the other hand we do
   need the special nature of the condensation to prove
   the converse in Exercise 6. Or at least we need the
   fact that every component is strong.

7. If the points of the factor digraph are numbered
    1,2,3,4 in the order given, the arcs of the factor
    digraph are (1,2), (1,3), (2,1), (3,1), (3,4), (4,1),
    (4,2), (4,3).

8. The complete list of components is $\{a,b,f,g,j,k\}$,
    $\{c,d,e,i\}$, $\{h\}$, $\{l,n,o\}$, $\{m\}$.

12. If $v \in R^O(u)$ then there is a walk
    $$<u = u_1,u_2,\ldots,u_m = v>$$
    from u to v. Then for each i, $\{u_i,u_{i+1}\}$ is an inner
    set of the digraph, and therefore a subset of $W(u)$.
    So $S(v) = \{u_1,u_2,u_{m-1},v\}$, being a union of subsets of
    $W(u)$ is itself a subset of $W(u)$. If these sets $S(v)$
    are found for each $v \in R^O(u)$, their union in $R^O(u)$
    and is also a subset of $W(u)$. So $R^O(u) \subseteq W(u)$.
        In the same way $R^I(u) \subseteq W(u)$.
    As $C(u)$ is a subset of $R^O(u)$, it is therefore a subset
    of $W(u)$ by the transitivity of containment.

16. There must be a closed walk through u and v. But if
    the walk passes through any other point, it also must
    belong to the component, hence the closed walk is
    $<u,v,u>$, and this requires both arcs (u,v), (v,u).

18. Point m is the source and p,q and t are the
    sinks. It is not true that there
    is always a walk from any source to any sink. In an
    acyclic digraph there must be a walk into each sink
    from some source.

19. The components are: $C = \{Ch,Hh\}$, $E = \{Ev\}$, $F = \{Fb,Fg\}$,
    $G = \{Go\}$, $H = \{Hg,Hm,Mk,Ly\}$, $S = \{St\}$.
    The arcs of the condensation are (C,H), (D,H), (F,H),
    (G,H), (S,H).

Exercises 2.11

1. $<1,2,3,4,8,7,12,PB,13,16,15,14,10,11,7,3,2,6,9,5,1>$
    is a spanning closed walk.

4. B and G are strong.

5. Taking any two points x,y there is an arc (x,y) and
    an arc (y,x), so there are walks $<x,y>$ and $<y,x>$.
    Hence any two points are equivalent.

6. Find any spanning closed walk in D. Reversing the
    order of the points gives a spanning closed walk in D^c.

7. The conjecture is false.

9. The minimum is p arcs.

10. Consider any digraph which is not strong. It
    therefore has at least two components. If each
    component subdigraph is not complete, arcs can be
    be added within the subdigraph without altering the
    digraph structure. Also so long as there remain at
    least two components, components can be amalgamated
    If C and D are two components all arcs beginning at a
    point in C, say, and ending in D can be present.
    Hence the maximum number of arcs will occur when there
    are two components C,D with m and n-m points
    respectively, each component subdigraph is complete,
    and there is an arc from each point in C to each point
    in D.
    Thus there are
    $$m(m-1) + (n-m)(n-m-1) + m(n-m) \text{ arcs.}$$
    This simplifies to
    $$(n^2-n) - m(n-m).$$
    The first term is fixed and the second will be least
    when m or n-m is 1. Thus the maximum number of arcs
    occurs when the two components have 1 and n-1 points
    respectively, and there are then $(n-1)^2$ arcs.

16. Insum 57, outsum 67, totsum 124.

17. On removing an arc no distance is decreased and at
    least one distance (that between the beginning and
    the end of the arc) is increased. Thus the sum of
    table entries is increased.

20. Incentre f, outcentre a, totcentre c.

CHAPTER 3.

Exercises 3.4

2. Only a can be 1; either f or g is 7. All precursors
    of d must precede it, all successors follow it, so d
    is at least 4 and at most 6. The smallest number a
    point can be given is its number of precursors, the
    largest number is
    (number of points) - (number of successors) + 1.

14. Scores only increase, so the scores are a logical
    numbering.

16. Part (i) follows from $<u>$ being the only path with u
    as last point.

16. (continued)
    Part (ii) needs to be proved in two parts. First,
    h(u) cannot exceed h(v) + 1, where v runs over all
    precursors of u, since there is some path of length
    h(u) into u, and the penultimate point of that path
    is the last point of a path of length h(u) - 1, and
    is also a precursor of u.
    Secondly h(u) cannot be less than h(v) + 1, for if
    W is any path into a precursor v of u, then W
    followed by <v,w>is a path of length one greater with
    u as last point. Hence for each precursor v of u,
$$h(u) > h(v).$$

17. You could apply a logical numbering and then proceed
    by induction on the logical number.

18. a:0, b:1, c:1, d:2, e:2, f:2, g:3.

20. One solution is:
    1. Convert the successor table to a precursor table.
    2. Assign level 0 to all points without precursors.
    3. Delete points to which a level has just been
       assigned wherever they occur as precursors.
    4. If there are no points to which no level has been
       assigned, end.
    5. Otherwise assign the next integer as level to all
       points which have not previously had a level
       assigned but now have no successors. Go to 3.

Exercises 3.8

1. The conjecture is false.

2. Arc (9,12) is the only inessential arc.

4. The key is that every walk in (V,A) is also a walk in
   (V,B).

5. Use Theorem 3.14 and Theorem 3.10.

17. Walnut 6/7; bread 3/7; wheatgerm 2/7; oatmeal 2/7;
    cheese 0; rice -2/7; lentils -4/7; millet -1.

18. All points in the same component get the same ranking
    coefficient.

19. We have to find a collection of walks such that each
    arc is in at least one walk. Seven walks seem to be
    necessary.

CHAPTER 4

Exercises 4.6

1. Spanning walks include: A: $\langle 1,2,3,4,5,6 \rangle$ (unique);
    B: $\langle 2,1,3,4,5,6 \rangle$ (unique);   C: $\langle 1,3,4,2,5 \rangle$;   E and
    G are also unilateral.

2. Writing (V,A) for the digraph and U for the defining
    subset of V, simple solutions include:

    (a) $V = \{1,2,3\}$, $A = \{(1,2),(2,3)\}$, $U = \{1,3\}$.
    (b) $V = \{1,2,3\}$, $A = \{(1,2),(1,3)\}$, $U = \{1,2\}$.
    (c) $V = \{1,2,3,4\}$, $A = \{(1,2),(2,3),(2,4),(4,2)\}$.

3. The core of the proof is that if $\langle u_1,u_2,\ldots,u_n \rangle$ is a
    spanning walk of a digraph then $\langle u_n,\ldots,u_2,u_1 \rangle$ is a
    spanning walk of the converse digraph.

6. For any u and v, either u is reachable from v, in
    which case
    $$R^O(u) \subseteq R^O(v),$$
    or v is reachable from u, in which case
    $$R^O(v) \subseteq R^O(u), \text{ by Theorem 2.1.}$$

7. PROOF. Suppose v and w are points both of indegree O
    in some digraph. As v has indegree O it is reachable
    from no other point, so it is not reachable from w.
    Similarly w is not reachable from v. Hence neither
    is reachable from the other and the digraph is not
    unilateral. The outdegree case is similar.

12. The properties of being complete, acyclic and
    symmetric are hereditary. The properties of being
    strong, weak, unilateral and self-converse are not
    hereditary.

14. Suppose T is a tournament with p points, in which no
    two points have the same outdegree. Arrange the
    points in order of decreasing outdegree. Then
    $$p-1 \geqslant od(u_1) > od(u_2) > od(u_2) \ldots > od(u_p) \geqslant 0$$
    But between p-1 and 0 inclusive there are only p
    values, so that
    $$od(u_1) = p-1, \quad od(u_2) = p-2,\ldots,od(u_p) = 0.$$

    Hence T is acyclic by Theorem 4.16, and therefore in
    a cyclic tournament some two points must have the same
    outdegree.

15. As each match involves two competitors, there can be
     at most ½p matches per day, and if p is odd this is
     reduced to ½(p-1). As the total number of matches is
     ½p(p-1), at least p days are required.

Exercises 4.9.

1.   Let x and y be the number of points in the two
     components. Then the numbers of arcs in the two
     components are ½x(x-1) and ½y(y-1). The total of arcs
     is 36, and x + y = 9. Hence

$$(x^2 + y^2) - (x + y) = 36.$$

     Putting y = 9 - x, this reduces to

$$(x - 3)(x - 6) = 0,$$

     so the cardinalities are 3 and 6, in either order.
     [It would actually be quicker to find the number of
     "internal" arcs for all possibilities in this case].

2.   The components are {New Zealand, Canada,
     United States}, {Japan}, {Taiwan}, {Guam}, {South
     Africa}. United States beat New Zealand and New
     Zealand beat Canada. All other results are that the
     team in the earlier component won.

4.   The components are {A}, {B,C,D}, {E}, {F,G,H,I}. The
     number of matches for which the result cannot be given
     is 0 + 3 + 0 + 6 = 9.

8.   Dominance analysis shows that {Apes, Bears, Cats} is
     a component, so the rest of the competition is
     irrelevant.
     Either: Apes beat Bears, Apes tied with Cats, Bears
             beat Cats
         or: Apes tied with Bears, Apes beat Cats, Bears
             tied with Cats.

CHAPTER 5

Exercises 5.3

1.   The tree is large if completely constructed, but may
     be greatly reduced by observing that if no 50c. coin
     is included the maximum possible total is 100c. from
     five 20c. coins, so that no other case without 50c.
     coins need be considered.
     The solutions are:

20 + 20 + 20 + 20 + 20;   50 + 20 + 10 + 10 + 10;
50 + 20 + 20 + 5 + 5;   50 + 20 + 20 + 10;   50 + 50.

4.   Start with Alan and Frances.   The tree is then as in
      Figure S.1, yielding four solutions.   (The question
      asks for adjacencies, so that the fact that each
      solution may be laid out either clockwise or anti-
      clockwise is irrelevant).

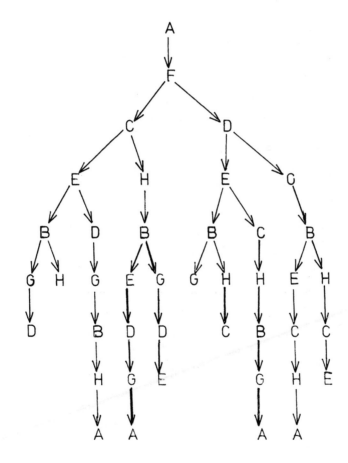

FIGURE S.1.

6. Consider the elements in order of decreasing atomic weight: S, Na, O, H. The final step is then always to insert however much hydrogen is required to make the total up to 120. There are nine solutions, of which $Na_1 H_1 S_1 O_4$ and $Na_1 H_1 S_2 O_2$ are the most plausible chemically.

8. There are eleven solutions:

R	R	B	B	P	P	P	P	P	P	P
G	W	G	W	G	G	G	W	W	W	P
P	P	P	P	R	G	P	R	G	P	G

9. The essence of the proof is to adapt the proof of Theorem 3.4. As there is only one source, this establishes a walk from the source to each point. Moreover the uniqueness of the walk is guaranteed by the fact that each point has a unique precursor.

10. Each point other than the root has one arc into it: this establishes a one-one correspondence between the arcs and the non-root points.

11. It appears that the minimum is $(n-1)$ when $t = 1$, $p = n - 1$ or $t = n - 1$, $p = 1$, and that the maximum is $\frac{1}{4}n^2$ when n is even and $\frac{1}{4}(n^2 - 1)$ when n is odd. These values are obtained when the root has $\frac{1}{2}n$ or $\frac{1}{2}(n + 1)$ successors respectively of which all but one are terminal and the remaining points form a 'chain' to a single terminal vertex.

13. (b) $R = T + (T^S + T^S)^S$

   (d) This can never be true, for in $(R + S)^S$ the root has outdegree 1, and in $R^S + S^S$ it has outdegree 2.

   (e) Yes. This can be proved by induction on the number of points, with $P(n)$ the statement that all rooted trees with at most n points can be so expressed. If the root has outdegree 1, the tree can be formed by stalking a rooted tree with n-1 points. Otherwise it can be formed by adding two trees each with less than n points.

Exercises 5.6

1. The shortest path is $\langle \alpha, 1, 2, 5, 10, 12, \omega \rangle$, with length 20.

2.   The shortest routes (cycles) are $<A,B,D,E,C,A>$ and
      its reverse $<A,C,E,D,B,A>$. Cycles occur in pairs of
      equal length because reversing the order does not alter
      the length.

3.   The best assignment is
            (A,Z), (B,Y), (C,V), (D,W), (E,X),
      with sum 37.

4.   First place the objects in the order 7,1,5,6,2,4,3.
      If the objects had been divisible the solution would
      have been 7,1,5,6 and one fifth of 2, value 258. The
      actual solution is 7,1,5 and 6, value 255, with 10
      units of volume unused.

5.   (a)   In order to force a win A must place the counter
            on either 6 or 7. If it is at 6 it will also be
            A's turn to move from 3: he must then choose 7.
            If he places the counter on 7 he must move to 4
            when the counter is on 3.

      (b)   A may again place the counter on 6 or 7, but
            taking now the other choice at 3. Or he may
            place it at 1 and make either choice at 3.

6.   This is a factor digraph, the sets of points being
      {L,P}, {K,M}, {N,Q}, {O,R}, {T,U,V,W}, with the rest
      singletons.

7.   The algorithm may be written as:

      1. Apply a logical numbering to the points.
      2. Consider point W with highest number not yet
          considered.
      3. If W is a sink, determine from the rules of the
          game the result represented by W. Go to 6.
      4. If W is not terminal identify the player whose
          moves are represented by the arcs out of W. Let
          this player be Y.
      5. If any arc out of W goes to a point labelled 'Y
          wins' label W 'Y wins" also, and label with a
          large arrowhead all arcs from W to such points.
          Otherwise label W to show that the other player
          wins, and put large arrowheads on all arcs out of
          W.
      6. If the source has been considered, go to 7.
          Otherwise go to 2.
      7. The winner will be the player whose label is on
          the source, and the natural play can be determined
          from the source by paths in the partial digraph
          consisting of the marked arcs.

CHAPTER 6.

Exercises 6.4

1. There are five such networks.

2. There are at least twenty, exactly half of which have the arc (α, ω).

4.

Point	α	1	2	3	4	5	6	7	8	9	10	11	12	13	14	ω
distance	0	3	4	8	9	14	6	19	32	11	12	51	10	16	23	17

5. 5,6,12,20,24,35.

6. List B contains the squares of integers up to 121.

13. Consider the points in order of some logical numbering. The value of the source is set at the maximum capacity of all arcs beginning there. The value of each other point is the maximum, over all arcs into the point, of the minimum of the capacity of the arc and the value of the point at the beginning of the arc. Record also the arcs for which this maximum is attained. The capacity of the maximum path is then the value of the sink, and the paths of maximal capacity are all paths from source to sink in the subdigraph consisting of all recorded arcs.

16. The numbers of paths from α to ω through v is $p(v).q(v)$. The number of paths from α to ω which contain (u,v) is $p(u).q(v)$.

17. First label the points of the network with a logical numbering. The point α must be 1. Set $p(\alpha) = 1$. Consider points in order of the numbering. The number $p(v)$ for each v is the sum of $p(u)$ over all precursors u of v. Thus the calculation may proceed from source to sink. The sink receives the last number in the logical numbering. To find $q(v)$ we set $q(\omega) = 1$ and considering points in the reverse order of the numbering, $q(u)$ is the sum of $q(v)$ over all successors v of u.

We may note that $p(\omega) = q(\alpha)$ and is the number of all paths from α to ω.

CHAPTER 7.

Exercises 7.5

2. See Figure S.2.

i.	T	ES	LS	EF	LF	TF	FF	
α	0	0	0	0	0	0	0	crit.
1	5	0	0	5	5	0	0	crit.
2	10	0	1	10	11	1	1	
3	8	0	6	8	14	6	0	
4	5	5	5	11	11	0	0	crit.
5	12	5	6	17	18	1	1	
6	7	11	11	18	18	0	0	crit.
7	4	8	14	12	18	6	6	
8	6	18	18	24	24	0	0	crit.
9	10	8	14	18	24	6	6	
ω	0	24	24	24	24	0	0	crit.

Critical path: $\langle \alpha,1,4,6,8,\omega \rangle$
Earliest completion: 24

3.

i.	T	ES	LS	EF	LF	TF	FF	
0	0	0	00	0	0	0	0	crit.
1	6	0	30	6	36	30	30	
2	4	0	0	4	4	0	0	crit.
3	5	4	24	9	29	20	0	
4	6	4	4	10	10	0	0	crit.
5	4	4	6	8	10	2	2	
6	3	9	29	12	32	20	20	
7	10	10	10	20	20	0	0	crit.
8	12	20	20	32	32	0	0	crit.
9	4	32	32	36	36	0	0	crit.
10	0	36	36	36	36	0	0	crit.

Critical path: $\langle 0,2,4,7,8,9,10 \rangle$
Earliest completion: 36

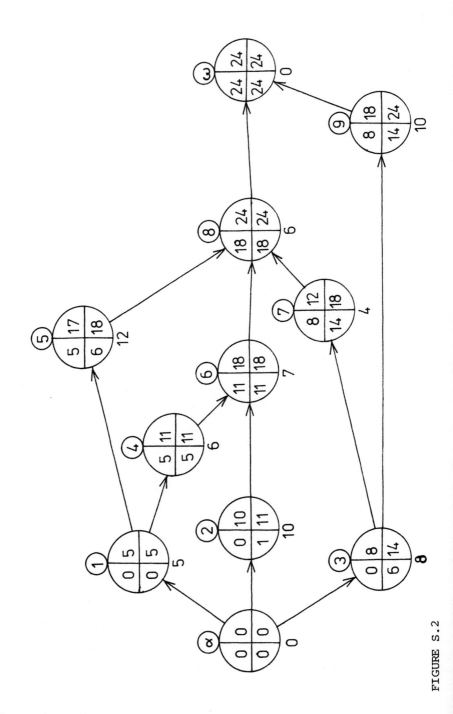

FIGURE S.2

4.   Activity 4 ceases to be critical: the new critical
      path is $<0,2,5,7,8,9,10>$, and the earliest completion
      time is now 34.

CHAPTER 8.

Exercises 8.5

1.   See Figure S.3.

(i)

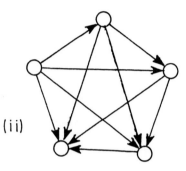

(ii)

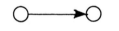

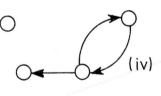

(iii)

(iv)

FIGURE S.3.

2.   (i) $\begin{bmatrix} 0 & 1 & 1 & 0 & 0 \\ 0 & 0 & 0 & 0 & 1 \\ 0 & 1 & 0 & 1 & 0 \\ 1 & 1 & 0 & 0 & 1 \\ 1 & 0 & 0 & 0 & 0 \end{bmatrix}$ (ii) $\begin{bmatrix} 0 & 1 & 0 & 0 \\ 1 & 0 & 0 & 0 \\ 0 & 0 & 0 & 0 \\ 0 & 0 & 1 & 0 \end{bmatrix}$

(iii)
$$\begin{bmatrix} 0 & 1 & 0 & 0 & 0 \\ 0 & 0 & 1 & 0 & 0 \\ 0 & 0 & 0 & 0 & 0 \\ 0 & 0 & 1 & 0 & 1 \\ 0 & 0 & 0 & 0 & 0 \end{bmatrix}$$
(iv)
$$\begin{bmatrix} 0 & 0 & 0 & 0 & 0 \\ 0 & 0 & 1 & 0 & 0 \\ 0 & 0 & 0 & 1 & 0 \\ 0 & 0 & 0 & 0 & 0 \\ 0 & 0 & 0 & 0 & 0 \end{bmatrix}$$

3. The digraph has 2 as source and 4 as sink.
   The cube of the adjacency matrix is

$$\begin{bmatrix} 0 & 0 & 1 & 1 \\ 1 & 0 & 1 & 2 \\ 1 & 0 & 0 & 1 \\ 0 & 0 & 0 & 0 \end{bmatrix}$$

As the $(2,4)$ entry is 2, there are two walks of length 3. (It is not necessary to calculate the whole matrix to find this number). The two walks are $<2,1,3,4>$ and $<2,3,1,4>$.

4. (i) is transitive but (ii) is not.

5. For (ii) the incidence matrix is

		arcs		
	1	2	3	4
1	1	0	0	-1
2	-1	1	0	0
3	0	-1	-1	0
4	0	0	1	1

points

7. As there is never more than one walk with given first and last points, the entries must all be 0 or 1. Further, for a particular (i,j) a_{ij}^n is 1 for at most one value of n, corresponding to the length of this path.

9. See Figure S.4.

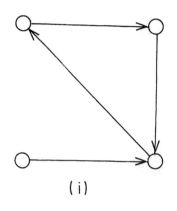

(i)

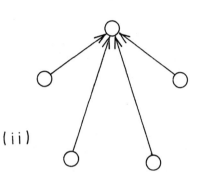

(ii)

FIGURE S.4

Section 8.8

Analysis of cycles and cut sets for the digraph in Figure 8.12 (i).

(a) The incidence matrix is

	1	2	3	4	5
1	1	0	0	-1	1
2	-1	1	0	0	0
3	0	-1	1	0	-1
4	0	0	-1	1	0

(b) The cycles are

     1: $\langle 1,2,3,4,1 \rangle$

     2: $\langle 1,2,3,1 \rangle$

     3: $\langle 1,3,4,1 \rangle$

    The cycle matrix is

	1	2	3	4	5
1	1	1	1	1	0
2	1	1	0	0	-1
3	0	0	1	1	1

(d) Every spanning tree will have three lines. The one used here has the lines corresponding to arcs 2, 4 and 5.

(e) The chords are 1 and 3 and the fundamental cycles are cycles 2 and 3 under (b).

$$F(D) \;=\; \begin{bmatrix} 1 & 0 & 0 & 0 & -1 \\ 0 & 0 & 1 & 1 & 1 \end{bmatrix}$$

(f) The remaining cycle 1 is the sum of cycles 2 and 3.

(g) Rearranging the arcs to put 1 first and 3 second, we obtain

$$\begin{array}{ccccc} 1 & 3 & 2 & 4 & 5 \end{array}$$
$$[\,I : \bar{F}(D)\,] \;=\; \begin{bmatrix} 1 & 0 & 1 & 0 & -1 \\ 0 & 1 & 0 & 1 & 1 \end{bmatrix}$$

(h) Removal of line 2 creates partition $\{1,3,4\},\{2\}$

4 $\{1,2,3\},\{4\}$

5 $\{1,4\},\{2,3\}$

	1	2	3	4	5
2	-1	1	0	0	0
4	0	0	-1	1	0
5	1	0	-1	0	1

There are six cut sets altogether; the other partitions being:

$\{1\},\{2,3,4\}$: row 5 - row 4

$\{3\},\{1,2,4\}$: row 2 + row 5

$\{1,2\},\{3,4\}$: row 2 - row 4 + row 5.

Reordering the columns as in (g), K becomes

$$[\bar{K}(D):I] = \begin{bmatrix} -1 & 0 & 1 & 0 & 0 \\ 0 & -1 & 0 & 1 & 0 \\ 1 & -1 & 0 & 0 & 1 \end{bmatrix}$$

Thus we confirm that $\bar{F}(D) = -K(D)^T$.

CHAPTER 9.

Exercises 9.4

3. The families and classes are each determined by
    disjoint complete subgraphs. The families are:
    {Timothy, Rebecca, Stephen}, {Euan, Alan, Heather},
    {Fiona, Nicholas}, {John}. The school classes are:
    {Timothy, John, Nicholas}, {Stephen, Fiona, Euan},
    {Rebecca, Heather}, {Alan}.

4. Any allowed combination will be a complete subgraph in
    the complement. In the restriction graph 113 is an
    isolated point, and may therefore be taken with any
    combination of mutually allowable courses.
    Enumeration by rooted tree is a suitable method for
    finding the maximum allowed combination. There are
    two maximum allowed combinations each
    worth 20 points: {103, 104, 110, 113},
    {103, 104, 112, 113}.

5. See Figure S.5.

8. If the rows and columns of the matrixes of G_1 and G_2
    are suitably ordered then the adjacency matrix of G_2
    is a submatrix of the adjacency matrix of G_1.

J

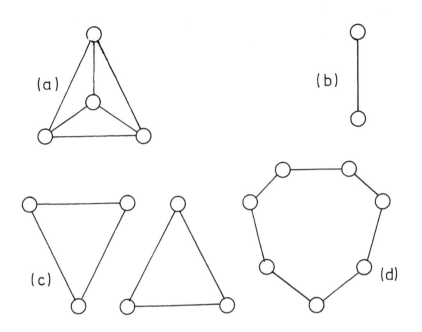

FIGURE S.5

9.   If the rows and columns of G_1 and G_2 are ordered in
      the same way, G_1 has a 1 whenever G_2 has a 1.

10.   The cliques are $\{1,2,3,4\}$, $\{1,2,4,5\}$, $\{1,5,6\}$, $\{2,5,7\}$,
      $\{5,6,7\}$, $\{7,8\}$.

13.   (a) is not self complementary since a self
          complementary graph on 4 points should have only
          3 lines.
      (b) is not self complementary since a self
          complementary graph on 5 points should have 5
          lines.

Exercises 9.7

2.   A graph which contains a cycle of odd length cannot be
      bipartite. This rules out (a), (c), (e) and (f).
      Both (b) and (d) are bipartite.

3.   One solution is to start from some arbitrary point u.
     Find the distance from u to each point v. Then the two
     sets are those points at odd distance from u and those
     points at an even distance. (This partition is in fact
     unique).

5.   These trees are the 'stars' $K_{m:1}$ for $m \geqslant 1$.

6.   For graphs insum and outsum coincide. A centre is a
     point of minimal sum. That there is a single centre or
     two adjacent centres follows from the result that if
     $\langle u,v,w \rangle$ is a walk then $s(u)$, $s(v)$, $s(w)$ are not all
     equal, and $s(v)$ is not the biggest.

7.   See Figure S.6. There are other ways in each case.

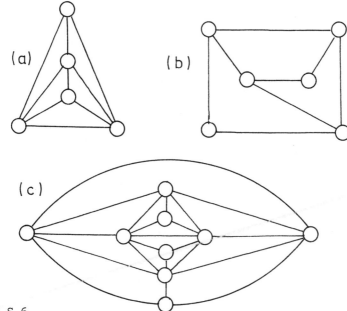

FIGURE S.6.

8.   One application is in <u>matching</u>. Suppose $W_1, \ldots, W_m$
     are workers and $J_1, \ldots, J_n$ are jobs. We can then
     set up a graph whose lines W_i, J_j imply that worker
     W_i is suitable for job J_n.

We then seek to pair workers and jobs so that as many positions as possible are filled. The <u>assignment problem</u>, (See Exercise 5.6.3) is a version in which the lines have weights corresponding to the suitability of W_i for J_j.

11. In Figure 9.7, (a), (b), (c), (e) are cacti: (c) is the only smooth cactus.

12. As each line is in exactly one cycle, the number of lines is $c_1 + c_2 + \ldots + c_k$

To find the number of points consider the cactus as 'growing'. All points of the first cycle count. Then a new cycle grows from some point of that cycle. This introduces (x-1) new points if the new cycle has x points. As each new cycle is added to the cactus already grown, the number of new points is one less than the length of the cycle. Thus the number of points is $(c_1 + c_2 + \ldots + c_k) - (k - 1)$.

13. If G is not connected let X,Y be two non-empty sets such that no edge in G joins a point in X and a point in Y. Then G' contains as a partial graph the complete bipartite graph on sets X and Y. Thus two points in different sets are adjacent, and two points in the same set are both adjacent to every point in the other.

19. Delete the line of greatest weight from each cycle.

20.

	a	b	c	d	e	f	g
Tree	No	I	Both	No	No	No	No
bipartite	Both	I	Both	Yes	No	No	Both
connected	Yes	I	Both	No	Both	Yes	No
self complementary	No	I	No	No	No	Yes	No

Yes: graph must have the property
No: graph cannot have the property
Both: there exist graphs with the property and graphs without.
I: No such graph is possible.

Some rules are:
(i)   (b) is impossible because the sum of degrees is odd.
(ii)   A tree must have $v \geqslant 2$ and
       $v_1 + 2v_2 + \ldots + kv_k = 2(v_1 + v_2 + \ldots + v_k - 1)$

(iii) In a self complementary graph not only must the numbers of points and lines fit the argument on page   but

$$v_k = v_{n-k-1}$$

for each point of degree k in G has degree n-k-1 in G'.

(iv) In f only the self-complementary graph can have these degree frequencies. Each point of degree 6 must be adjacent to each of the other points except one. These must be pendant points in each case, for if either point of degree 6 is a adjacent to both pendant points, the other is adjacent to neither and so cannot be adjacent to six points.

(v) A graph in which v_o is not zero cannot be connected (unless it is the graph with one point).

22. (a) Omit the line of weight 3, and the line of weight 6 on the left. Omit either (but not both) of the lines of weight 6 on the right.

(b) Omit the lines of weights 40, 19, 7 and 25.

(c) This is its own spanning tree of minimal weight.

(d) Choose the lines of weight 25, 50, 60, 75, 100 and 400.

(e) Omit the line of weight 43.

(f) A minimal spanning tree consists of any three lines which do not form a triangle.

FURTHER READING

There are a few works which deal solely with digraphs, but most deal with both graphs and digraphs, the emphasis on the former. (The books below are referred to at various places in the main text thus: [Kaufmann].)

A) Works dealing solely with digraphs:

    1. Kaufmann, A., Points and Arrows, Transworld Library, London, 1972.
This is a nonspecialist booklet with a number of interesting applications. Written for the intelligent layman in relatively nontechnical language, it introduces some topics of digraphs in an interesting way.

    2. Harary, F., Norman, R.Z., Cartwright, D., Structural Models: an introduction to the theory of directed graphs. Wiley, New York, 1965.
This is a textbook which presents the theory of digraphs with a view to applying it to psychology.

B) Tournaments:

    3. Moon, J., Topics on Tournaments; Holt, Rinehart, and Winston, New York, 1968.
A monograph on tournaments. Recommended to the student who has read chapter 4 and wishes to learn more about tournaments.

C) Applications to Operations Research:

For a detailed description of the solution technique for the burglar's problem see:

    4. Kolesar, P.J., "A Branch and Bound Algorithm for the Knapsack Problem," Management Science, v.13, (May, 1967) no. 9, pp.723.

    5. Lawler, E., L., Wood, D.E., "Branch and Bound Methods: A Survey," Operations Research v.14 (1966) No. 4 pp.699-719.
This article requires a reasonable level of mathematical maturity. It describes the essential features of branch and bound. It also contains applications of the technique to various problems in operations research.

6. Plane, D.R., MacMillan, C., Discrete Optimization, Prentice-Hall, (1971).
    This book has an undergraduate approach to many discrete problems in O.R. including: branch and bound enumeration, shortest path problems, and minimal spanning trees. It does not assume much mathematical knowledge.

7. Taha, H., Operations Research: An Introduction, Macmillan.
    An excellent introduction to O.R. It covers many of the topics mentioned in chapters 5,6, and 7. It assumes knowledge of the differential calculus and linear algebra.

D) Activity Networks:

The following works refer to the activity-point approach as described in chapter 7.

8. Armstrong-Wright, A.T., Critical Path Method, Longmans, 1969 (chapter 15).

9. Archibald, R.D., Villoria, R.L., Network based Management Systems, Wiley, New York, 1967. Appendix B.

10. Burman, P.J., Precedence Networks for project planning and control, McGraw-Hill, 1972.

11. Larkin, J.A. Card Network Planning, Scottish Academic Press, 1970.

12. Moder, J.J. Phillips, C.R., Project Management with C.P.M. and Pert, Reinhold, 1970, (Appendix 2-1).

13. Owen, G., Finite Mathematics, Saunders, 1970, (Ch.8 section 2).

E) Graphs:

The following texts deal with both graphs and digraphs:

14. Berge, C., The Theory of Graphs and its Application Applications, Methuen/Wiley, 1962.

15. Busacker, R.G., Saaty, T.L., Finite Graphs and Networks, McGraw-Hill, 1970.

16. Chen Wai-Kai, Applied Graph Theory, North Holland, 1971.

F) The following texts refer to digraphs but place their major emphasis on graphs:

17. Bondy, J.A., Murty, U.S.R., Graph Theory with Applications, Macmillan, 1976.
    The chapter on digraphs introduces paths and cycles

and contains applications to job sequencing, efficient computer drum design, road systems, and tournaments.

18. Anderson, S.S., Graph Theory and Finite Combinatorics, Markham, 1970.

19. Deo, N., Graph Theory with Applications, Prentice Hall, 1974. This text contains an excellent chapter on digraphs and many applications to operations research.

20. Harary, F., Graph Theory, Addison-Wesley, Reading, Massachussetts, 1969.
This is the definitive work on graph theory by a leader in the field. It contains a chapter on digraphs with applications to tournaments. It also lists all digraphs with less than five points.

21. Behzad, M., and Chartrand, G., Introduction to the Theory of Graphs, Allyn and Bacon, 1971.

22. Marshall, C.W., Applied Graph Theory, Wiley-Interscience, 1971.

23. Ore, O., Graphs and Their Uses, Random House, 1963.

24. Wilson, R., Graph Theory. Oliver and Boyd, 1972.

INDEX

Underlined page numbers indicate a
definition, either formal or informal, of the term.

active 127
activity 151
activity arc approach 165
acyclic digraph 55,70,175
acyclic tournament 105
addition of trees 122
adjacency matrix 169
adjacent 19,169,202
algorithm 48,158
antireflexive 14
arc 18
arrow diagram 14
assignment problem 134
automorphism 40
beginning 19
bigate 54
bipartite 213
block diagonal 178
bound 130
branch 189,201
branch and bound 124
branching 130
burglar problem 128
cactus 222
carbon cycle 34
cardinality 3
cartesian product 3
central 66
chord 189,218
clique 210
closed walk 25,214
column of matrix 5,170
column vector 6
compactness ratio 67
complement 2,205
complementary 205
complete bipartite
 graph 213

complete digraph 20
complete graph 205
component 47,50,214
component analysis 48
component subdigraph 50
condensation 53
connected 214
contained in 2
contains 2
contrapositive 30
converse 36
course prerequisite 45
cost 135,140
critical activity 151,157,
 160
critical path 157,160
critical path method 165
cumulative deficit 110
cut set 193
cycle 25,182,214
cycle matrix 184
cyclic digraph 70
deficit 110
degree 202
diameter 65,214
diametral path 65
digraph 16,18
dimension 6
directed edge 18
directed graph 18
disorientation 182
distance 26,64,214
division 14
domain 14
dominance 112
dummy activity 165
duration time 152
earliest finish time 157

earliest start time   157
edge   201
element   1
empty set   2
end   19
entry   5
enumeration   116,124
equal digraphs   18
equal matrices   6
equal ordered pairs   3
equal sets   2
equivalence class   16
equivalence (digraphs)   46,56
equivalence relation   16,40
essential arc   82,85
essential part   84
event   153
exhaustive enumeration   124
factor digraph   55
fathoming   126
feasible solution   126,129
finite set   3
first point   24
float   157,161
flowchart   74,76,144
food preferences   94,96,107
food web   22
forest   215
four colour problem   212
free float   163
function   3,18
fundamental cut set   193
fundamental cut set
 matrix   195
fundamental cycle   187
fundamental cycle matrix   191
games   132
graph   38,173,202
hereditary   107
hierarchy   123
home and away matches   105
identity matrix   9
implicit enumeration   124
incentre   66
incidence matrix   176
incident   176,202
incumbent   126

indegree   20,170
induction   11
inessential arc   82,85
infinite set   3
ingate   54
inner product   7
inner set   59
inreach   44
insum   66
intersection   1
inverse function   4
invertible   4
isolated point   20,171
isomorphism (digraphs)   39
isomorphism (graphs)   203
isomorphism invariant   40
join   202
join of walks   26
kitset instructions   77
Kruskal's method   220
last point   24
latest finish time   157
latest start time   158
leading diagonal   9
length   26,28,157,214
lesson preparation   77
Lewis Carroll   29
levels   79
levels function   79
line   38,201,202
linearisation   91
logic   29
logical numbering   72,75
loop   16
matching   247
mathematical induction   11
matrix   5
matrix product   7
matrix sum   6
matrix transpose   10
maximum   11
minimally connected   216
minimum   10
multigraph   182
network   18,138,140
node   18
noncritical   157

nonlinear ranking	86	scalar product	7
nonplanar	212	section of digraph	57
null set	2	self-complementary	206
objective function	126	self-converse	42
one-one correspondence	4	set	1
ordered pair	2	shortest path	116,126,140
orientation	182	shortest path algorithm	142
outcentre	66	sigma	4
outdegree	19	sink	20,170
outer set	59	smooth cactus	222
outgate	53	sociogram	52
outreach	44	source	20,170
outsum	66,170	spanning cycle	26
paired comparison	96	spanning path	96,105
partial digraph	35	spanning tree	189,218
partial graph	203	spanning walk	26
partial ordering	87	splitting a walk	27
partial path	126	square matrix	5
partial subdigraph	35	star	247
partition	17	strong	26,55,63
path	25,214	strong component	59
penalty function	127	strong tournament	102
pendant point	218	strongly connected	59
planar graph	212	subcomponent	57
planarity	212	subdigraph	35
point	18,202	subgraph	203
precedence	77,152	subset	2,87
precursor	19	subtournament	102
proper cut set	193	successor	19
proper sink	20,171	successor table	19
proper source	20,171	summation	4,7
propositions	29	supercomponent	57
pruning	127	supergraph	203
pseudodigraph	171	superset	2
Pythagorean mathematics	121	symmetric digraph	37,173,202
ranking coefficient	91	symmetric matrix	10,174
reachable	44	symmetric relation	16
reflexive relation	14,16	ties	112
relation	13	total float	161
reverse	36	totcentre	66
root	115	totsum	66
rooted tree	115	tournament	95,97
round robin	95	transitive closure	87
row of matrix	5,170	transitive digraph	87,174, 175
row vector	6		
rugby football	32,96	transitive relation	16
scalar multiple	6	transport network	139

transpose	10	value	$\underline{4}$
travelling salesman		vector	6
problem	134	vertex	18
tree	$\underline{215}$	walk	$\underline{24},\underline{214}$
triangle	$\underline{205}$	weak	$\underline{59}$
triangle inequality	$\underline{28}$	weak component	$\underline{59}$
trivial walk	$\underline{25}$	weakly connected	$\underline{59},192$
unilateral digraph	$\underline{98}$	weight	219
union	$\underline{2}$	wins analysis	110
unit matrix	$\underline{9}$	zero matrix	$\underline{6}$
upper triangular form	176		